INTRODUCTORY CHEMISTRY LABORATORY MANUAL

Second Edition

VAN-GRINER LEARNING

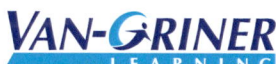

Introductory Chemistry

Laboratory Manual
University North Carolina Greensboro
Second Edition

Copyright © by Spencer Russell
Copyright © by Van-Griner, LLC

Photos and other illustrations are owned by Van-Griner Learning or used under license.
All products used herein are for identification purposes only, and may be trademarks or registered trademarks of their respective owners.

All rights reserved. No part of this book may be reproduced or transmitted in any form or by any means, electronic or mechanical, including photocopying, recording or by any information storage and retrieval system, without written permission from the author and publisher.

The information and material contained in this manual are provided "as is," and without warranty of any kind, expressed or implied, including without limitation any warranty concerning accuracy, adequacy, or completeness of such information. Neither the authors, the publisher nor any copyright holder shall be responsible for any claims, attributable errors, omissions, or other inaccuracies contained in this manual. Nor shall they be liable for damages of any type, including but not limited to, direct, indirect, special, incidental, or consequential damages arising out of or relating to the use of such material or information.

These experiments are designed to be used in college or university level laboratory courses, and should not be conducted unless there is an appropriate level of supervision, safety training, personal protective equipment and other safety facilities available for users. The publisher and authors believe that the lab experiments described in this manual, when conducted in conformity with the safety precautions described herein and according to appropriate laboratory safety procedures, are reasonably safe for students for whom this manual is directed. Nonetheless, many of the experiments are accompanied by some degree of risk, including human error, the failure or misuse of laboratory or electrical equipment, mis-measurement, spills of chemicals, and exposure to sharp objects, heat, blood, body fluids or other liquids. The publisher and authors disclaim any liability arising from such risks in connection with any of the experiments in the manual. Any users of this manual assume responsibility and risk associated with conducting any of the experiments set forth herein. If students have questions or problems with materials, procedures, or instructions on any experiment, they should always ask their instructor for immediate help before proceeding.

Printed in the United States of America
10 9 8 7 6 5
ISBN: 978-1-64565-205-2

Van-Griner Learning
Cincinnati, Ohio
www.van-griner.com

President: Dreis Van Landuyt
Senior Project Manager: Maria Walterbusch
Customer Care Lead: Lauren Wendel

Russel 65-205-2 F22
327759-333574
Copyright © 2024

TABLE OF CONTENTS

PREFACE V

LABORATORIES

- **LAB 1** TREATMENT OF SCIENTIFIC DATA 1
- **LAB 2** THINK METRIC 7
- **LAB 3** SUGAR CONTENT OF SOFT DRINKS 19
- **LAB 4** THE ENERGY CONTENT OF FOOD 27
- **LAB 5** SEPARATION OF A MIXTURE 33
- **LAB 6** FLAME TESTS AND PROPERTIES OF IONIC COMPOUNDS 43
- **LAB 7** LAW OF DEFINITE COMPOSITION 53
- **LAB 8** THE PREPARATION OF ALUM FROM SCRAP ALUMINUM 59
- **LAB 9** SAPONIFICATION OF VEGETABLE OIL AND SOAP PROPERTIES 67
- **LAB 10** SYNTHESIS OF ASPIRIN 77
- **LAB 11** IDEAL GAS LAW: MOLECULAR WEIGHT OF A VAPOR 85
- **LAB 12** SOLUTIONS AND SOLUBILITY 93
- **LAB 13** ACIDS, BASES, AND BUFFERS 107
- **LAB 14** DETERMINING THE PURITY OF ASPIRIN BY TITRATION 119

PREFACE

OBJECTIVES

During this course, the students will:

1. Learn and practice safety standards in a chemical laboratory environment;
2. Develop skills in handling scientific equipment and making measurements;
3. Apply record keeping and reporting methods that reflect the integrity and ethics associated with scientific data and information;
4. Apply scientific inquiry to develop knowledge of chemical systems;
5. Classify substances based on their physical and chemical properties;
6. Synthesize and characterize chemical compounds;
7. Perform calculations describing mass and energy changes during chemical reactions; and
8. Use scientific instruments to make quantitative measurements of the properties of samples.

INTRODUCTION

Science is a process of developing new knowledge that expands our understanding of the facts, theories, and laws that describe the natural world. By observing, formulating hypotheses, and testing, we achieve a deeper understanding of the world around us. Your experience in the science laboratory should provide you with an opportunity to explore the world of chemistry, to evaluate its theories and laws, and to solve problems associated with chemical systems. In addition, you will learn the methodology by which scientists control and manipulate chemical systems. Careful scientific work like this is the foundation for validating new knowledge that can be tested and confirmed by others.

You are encouraged to be observant and questioning. Enjoy the wonder of the chemical transformations that you will see and note the precise mathematical relationships that are part of our descriptions for chemical systems. Hopefully, your experience will be fulfilling and rewarding, and will expand your perspective of the chemical world around you.

INTRODUCTION

COMMON LAB EQUIPMENT

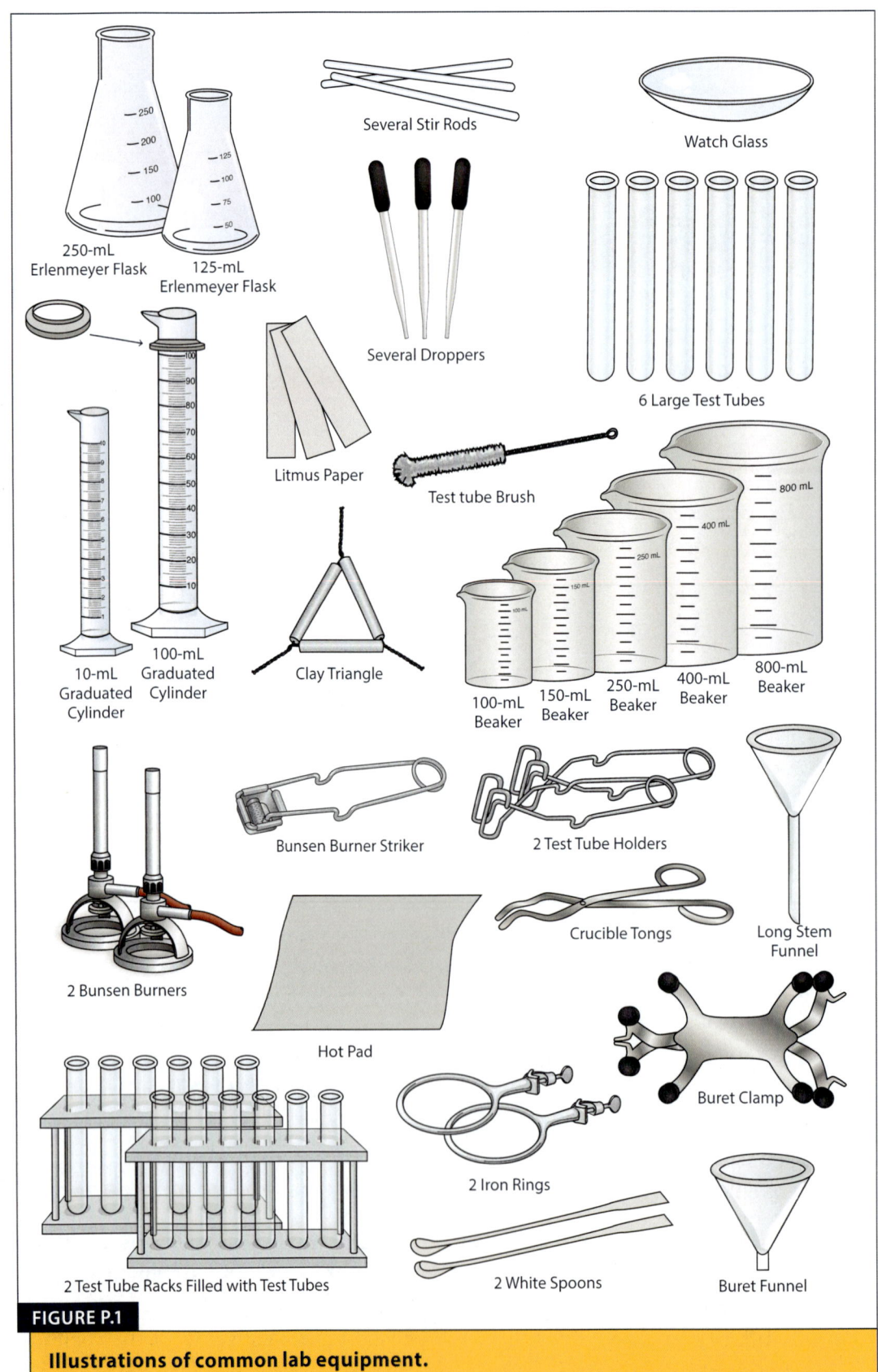

FIGURE P.1

Illustrations of common lab equipment.

STUDENT EXPECTATIONS

Good laboratory work consists of much more than just the mechanical procedure for carrying out experiments. It also requires thorough understanding of the experiment, planning for the work to be carried out, consideration for the safety and convenience of others while in the laboratory, and an accurate and clear record of all the information obtained during the experiment. Students who are successful in the course come prepared, follow directions, and adhere to policies.

1. **Attendance**

   You are required to attend all scheduled laboratory meetings. Excused absences are those involving a death in the immediate family or an illness supported by a doctor's note. For other situations, special arrangements should be made through the instructor prior to the class if at all possible. On rare occasions, you may be permitted to attend an alternative lab section during a given week. Arrangements with instructors in your regular lab and the lab you will visit should be made. **If you miss three laboratory meetings or fail to turn in three or more lab reports, you are likely to receive an F in the course (if unexcused) or be given an incomplete (if excused) because of the amount of work missed.** *Students who do not attend the prelab (or come late) will NOT be permitted to perform the experiment, and will receive a zero for that experiment.*

2. **Organization**

   You will be expected to prepare for each experiment before coming to the lab each week. Read the experiment to become aware of the important concepts and to get an overview of the procedure. In some cases, the experiment may include a problem or equations to be worked out beforehand.

   The lab period will be adequate for carrying out experiments and cleaning up afterwards if your work is planned before you come. The instructor will expect you to leave the lab on time. You will be sharing many pieces of equipment and reagents with other students, and you may have to wait your turn to use them. If your work has been planned properly, you can work on another phase of the experiment while you are waiting for a particular item, and therefore, you will make better use of your time.

3. **Safety**

   Your safety and that of others around you is the most important factor in laboratory work. The experiments have been found to be quite safe as long as you *follow directions.* A little thought on your part can prevent many careless accidents. For example, remember that an iron ring used as a support above a burner for heating water will remain very hot for some time. Contact with a liquid acid spilled on the bench top can result in serious burns. Be sure to wipe up spills in the proper manner. Flammable reagents or clothing should not be used near a flame. Chemical splash goggles must be worn at all times in the laboratory.

INTRODUCTION

Goggles must be worn at all times in the laboratory. Non-porous close-toed shoes covering the entire top of the foot must be worn at all times in the lab. Tights do not provide adequate protection from chemical spills and are not acceptable clothing. Pants are required while working in the laboratory. Failure to follow these rules can result in a zero for the day.

4. **Consideration of others**

    In sharing lab facilities with others, it is necessary to leave reagents and special apparatus on the shelves rather than removing them to your own desk for use. For example, to obtain 10 mL of a solution, take your graduated cylinder to the shelf and pour the solution there. Do not take the bottle back to your desk. If any material spills on the shelf or on the sides of the bottle, you should clean it up at once. Return bottles to their original location so that others may find them easily. Cap the bottles and close containers after each use.

    Before leaving at the end of the lab period, you are responsible for cleaning up the sink, your working area, and any equipment you have used. Carefully check to be sure that you have returned any special apparatus loaned for the lab period and that only the proper items have been replaced in your drawer.

    Any items that are found by the instructor at the end of the lab period will be returned to the stock supply and you will have to replace them at your expense.

    Trash should be thrown in the trash cans and not in the sinks, in the broken glass box, or on the floor.

    Cleaning up and restoring the lab to the condition in which you found it are considered part of your laboratory assignment. You will be graded on this aspect of your lab work.

5. **Lab write-ups**

    In the scientific world, great legal, moral, and ethical value is placed on the integrity of scientific information. Such integrity and honesty is fundamental to the advancement of scientific knowledge. In scientific and technical industries, the intellectual property rights (patents) are based on proper scientific records.

    The student in a science course, being a science investigator, has the same objective. Therefore, it is only logical that one should keep a permanent and complete record of the procedure followed and the data obtained. While someone else may have worked out the experimental procedure previously, the student is carrying out the scientific investigation for the first time and the information obtained has the same fundamental significance.

    All data, calculations, and results must be recorded on the Report Pages in this lab manual IN INK AT THE TIME THE INFORMATION IS OBTAINED DURING THE EXPERIMENT. You are to record all observations and data as you go along. No data are to be noted on loose slips of paper. When corrections are necessary, do not erase the error, but draw a single line (single line) through it and correct it nearby. No "white-out" is allowed. Each day's lab work must be dated and pages should be kept in order for later reference by you and your instructor. It is far more important to have a useful record containing a few smudges and errors, but which is kept up to date while you work, than a few absolutely neat—and artificial—pages.

Always include units of measurement and label each item clearly.

6. **Lab reports**

    You will be required to turn in a report of the observations, data, results, and conclusions for each experiment at the time specified by your instructor. The report will usually consist of the Report Page with your data, observations, calculated results, graphs, computer printouts, answers to questions, and your own individual comments. The Report Page usually contains questions that require you to relate the investigation to everyday situations. When presenting calculations, always write out the relationship to be used, show how values are entered, and then show the solution. Present answers to discussion questions in clear, complete sentences. **All of the work on the lab report must be your own, even if experiments are performed with partners or in groups. You are encouraged to discuss the information, but the written report must be your own work. Reports are entirely open for discussion until you are completing the Report Page. Completion of the Report Page is an open notes, open book exercise to be completed by you individually.** *Reports that show evidence of copying will lead to processing as Academic Integrity violations.*

    As often as time allows, your final results will be checked individually by the lab instructor before you leave. This has many advantages, including the fact that you will be able to consider the significance and validity of your results and discuss them with the instructor while you are still very familiar with the procedure and principles of the experiment. In instances where the procedure and calculations are especially lengthy, the instructor may allow you to complete the assignment outside of class and turn it in at a specified time. When the corrected report is returned to you, you should retain it in its proper place in your lab manual.

    Lab reports validate your completion of the experiment. If you do not turn in a report, it is assumed that you did not do the experiment and you receive a 0 for the experiment. Late reports will not be accepted and three zeros on lab reports will result in an F for the final grade in the course.

INTRODUCTION

7. **Clean glassware**

   Wash the apparatus as soon as possible after using. Materials are often harder to remove after standing. After rinsing with tap water, rinse again with a stream of deionized water from your wash bottle. Detergent solutions for washing glassware are available at the sinks. Do not use a large excess of detergent. Aside from being wasteful, it is hard to rinse the detergent off.

   Never use a paper towel to dry the inside of your equipment as the towel only contaminates it again. If the glass has been washed and rinsed properly, water will drain completely, leaving the surface dry—and very clean.

8. **Reagent bottles**

   When using reagent bottles, a little care will keep them clean and uncontaminated. On opening the bottle, be sure that the inside of the stopper cannot touch any foreign material. Be sure to close the cap of the bottle properly when you finish withdrawing the reagent. If any reagent runs down the side of the bottle, rinse the latter with water before placing it back on the shelf.

   Another good habit to acquire is that of pouring liquids from the side of the bottle away from the label. If everyone does this, the label will not be destroyed by spilled reagent and your hand will not come in contact with the spills which someone else may have forgotten to wash off.

   Always recheck each label before pouring to be certain you've picked up the right bottle, and, for a margin of safety, recheck the label, particularly the concentration of solutions, as you place the bottle back on the shelf.

   Never put anything into a reagent bottle! This includes droppers, spoons, fingers, spatulas, excess reagent, or anything else. To remove a solid material from the bottle, shake some out onto a piece of creased weighing paper, onto a watch glass, or into a beaker, using one of the approved methods suggested by the instructor. If the solid is lumped together, tap the outside of the bottle to loosen its contents. Frequently, a spatula or spoon will be provided in a beaker next to a reagent bottle specifically to help remove that one reagent.

   To obtain any amount of liquid reagent, pour it out into a container. If only a few drops are needed, you may then put a medicine dropper into the liquid you have poured out, but never into the original bottle. If the reagent bottle contains a dropper, do not let the dropper touch the side of the vessel into which you are transferring the reagent. In pouring liquids, you can prevent spilling them on the outside of the bottle by pouring the liquid down a stirring rod and finally touching the lip of the bottle to the rod, stopper, or receiving vessel to catch the last drop. If you pour out too much reagent, either throw away the excess or put it into the special container provided. *Never return it to the reagent bottle.* You should, of course, endeavor to pour out close to the amount of reagent needed to avoid needless waste of materials. If you obtain too much reagent, share it with a colleague.

9. **Other**

   Other suggestions for lab technique are given in individual directions for experiments or will be demonstrated by your instructor. Do not hesitate to ask instructors or assistants for any help. After all, you are here to learn.

   The development of good lab technique is only one of the functions of the lab sessions included in chemistry courses. An equally important purpose is to give you the opportunity to observe firsthand how matter behaves and how scientific laws and principles are developed. Therefore, it is essential that you observe and report what is actually going on during an experiment rather than try to make the results come out as you think they should. Don't worry if your results do not check with those of your neighbors. They may all be wrong and your results may be the accurate ones.

SAFETY, EQUIPMENT, AND EXPECTATIONS

LEARNING OBJECTIVES

1. To become familiar with the proper safety precautions and safety procedures in the laboratory
2. To locate the safety equipment in the laboratory
3. To commit to safety while working in the laboratory

INTRODUCTION

As in all laboratory settings, **SAFETY is paramount** in this course. Please read carefully and follow all the safety rules listed below at all times you are in the laboratory. The well-being of both you and your classmates depends on it.

Important Note: *Failure to abide by any of the rules below may result in your immediate removal from the laboratory with the assignment of a zero for the day's work.*

1. **Injuries in the lab.** Report any personal injury immediately to your lab instructor. Campus police should be the first contact for a serious incident. **Campus Police: 334-4444**

2. **Eye protection is required at all times.** North Carolina State Law requires students and staff to wear eye protection in the laboratory. This means eye covering which will protect both against impact and splashes. **Thus, *chemical splash goggles* are required at all times in the lab room.** Goggles must be worn even after you complete the experiment and must be worn until you are ready to walk out of the lab room. If you should get a chemical in your eye, wash with flowing water from an eye wash station for 15–20

INTRODUCTION

minutes. If you are wearing contact lenses, remove them immediately. (Data suggests that contact lenses present no unusual hazard as long as splash goggles are used for eye protection.) **Safety glasses are not a replacement for goggles.**

3. **Wear proper clothing.** Wear old clothes, long pants, and suitable footwear. **Shoes must cover your ENTIRE foot.** Open-toed shoes, sandals, ballet flats, porous shoes, or any footwear that does not cover the entire upper surface of the foot presents a particular hazard in the lab because of the foot's vulnerability to having chemicals splash on it or broken glassware fall on it. Therefore, these types of footwear are prohibited in the lab.

   Pants are mandatory in the laboratory. Tights are not adequate covering for the skin. Loose clothing of any kind also represents a hazard in the laboratory. Also, confine long hair when in the laboratory.

   Students who do not have goggles, proper footwear, and/or clothing will not be permitted to work in the lab, and they will receive a zero for their grade for that experiment.

4. **Know the location of all safety equipment.** Fire extinguishers, eye wash stations, safety showers, and the fire blanket are just inside the door to the lab. Note their location when you first enter the lab room. Review the usage of the eye wash station and safety shower with your lab instructor. In case of a fire, inform your instructor at once.

5. **No food or drink is allowed in the laboratory.** This includes water bottles.

6. **Clean up chemical spills and broken glass immediately.** Dispose of broken glass and waste chemicals in the appropriate waste containers.

7. **Never put any chemical in or near your mouth.** Never mouth pipet a solution. Also, exercise great care in noting the odors of fumes and avoid breathing fumes of any kind.

8. *Do not* **sit on the bench tops or the floor of the laboratory.** Chemical residues and small fragments of broken glassware make these practices hazardous to both clothing and personal health.

9. **Chemical hazard rating.** The general hazards of a chemical are typically presented on an NFPA label in a spatial arrangement of numbers with the flammability rating at the twelve o'clock position (usually red), the reactivity rating at the three o'clock position (usually yellow), and the health rating at the nine o'clock position (usually blue). At the six o'clock position, information may be given on special hazards such as W for reactivity with water, COR for corrosive properties, or OXY for oxidizing properties. A chemical is assigned a relative hazard rating that ranges from

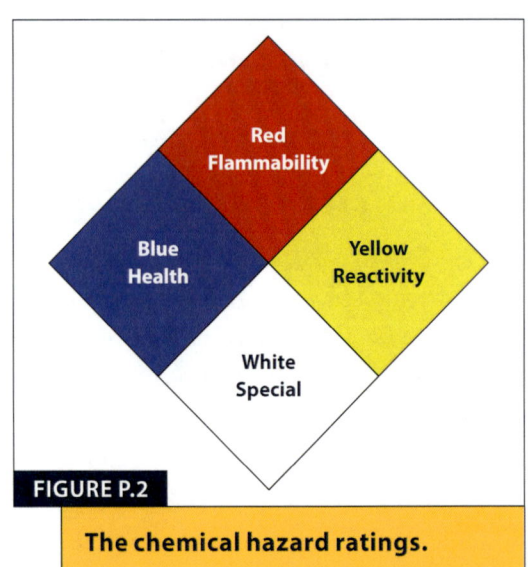

FIGURE P.2 The chemical hazard ratings.

XII INTRODUCTORY CHEMISTRY LABORATORY MANUAL

0 (minimal hazard) to 4 (extreme hazard). The health hazard indicates the likelihood that a material will cause injury due to exposure by contact, inhalation, or ingestion. The flammability hazard indicates the potential for burning. The reactivity hazard indicates the instability of the material by itself or with water with subsequent release of energy.

10. **Waste disposal.** You will produce chemical wastes in the laboratory. Although you will use small quantities of materials, some waste products are unavoidable. In order to dispose of these chemical wastes safely, you need to know some general rules for proper chemical waste disposal.

    a. **Metals.** Metals are to be placed in a container to be recycled.

    b. **Nonhazardous chemical wastes.** Substances such as sodium chloride (NaCl) that are water soluble and are not hazardous may be emptied into a sink. If the waste is a nonhazardous solid, dissolve it in water before disposal.

    c. **Hazardous chemical wastes.** If a substance is hazardous or not soluble in water, it must be placed in a container that is labeled for hazardous waste. Your instructor will inform you if chemical wastes are hazardous and identify the proper waste container(s). If you are unsure about the proper disposal of a substance, ask your instructor. The label on a waste container should indicate hazardous contents and the name(s) of the chemical waste(s).

11. **Perform no unauthorized experiments.**

12. **Never work alone in the laboratory.**

13. **Students who are pregnant are discouraged from taking the lab course. If you choose to stay in the course, a waiver formed must be signed.**

14. **Wash your hands and clean up your work area before you leave the lab.**

STANDARD LABORATORY PRACTICES

1. **Preparation.** Before you come to the laboratory, read the discussion of and directions for the experiment you will be conducting. Make sure that you know what the experiment is about before you start the actual work. Studying the purpose of an experiment is the most important step in chemistry laboratory practice.

2. **Check labels twice.** Be sure you take the correct chemical. Read reagent bottles carefully. There is a great difference between potassium chloride and potassium chlorate, between magnesium and manganese.

3. **Never directly pipet a solution out of a community stock container.** Always use a beaker or a graduated cylinder to take the appropriate volume back to your work area.

4. **Never take a community stock container to your bench.**

5. **Be conscientious regarding the amount of material you remove from community stock containers.** Determine the approximate amount needed (volume or mass) prior to removing any substance from a stock container to minimize chemical waste.

INTRODUCTION

6. **Never return unused portions of stock solutions to the community containers.** Share it with a colleague or dispose of it properly.

7. **Handling of equipment.** Test tubes or any pieces of equipment which have a potential for expelling a gas or other material should be pointed away from others in the laboratory and yourself.

8. **Check glassware.** Check all glassware for chips or cracks prior to use. Replace chipped or cracked glassware immediately.

9. **Flush drains.** Thoroughly flush drains after pouring out liquid reagents. Do not discard solids in a sink.

10. **Good housekeeping practices.** The continuous practice of good housekeeping is essential to the prevention of accidents, fires, and personal injuries. Each laboratory worker is responsible for the following:

    a. Keeping benches, tables, hoods, floors, aisles, and desks clear of all materials not being used.

    b. Keeping an adequate passageway clear to exits.

    c. Keeping clear space around eye wash stations, fire extinguishers, fire blankets, safety showers, and electrical controls.

    d. Keeping floors free of spilled water, dropped stirring rods, stoppers, pencils, and other slipping hazards.

    e. Cleaning up spills and disposing of the materials used to absorb the spills.

    f. Properly disposing of all chemical wastes generated. Make sure to follow your lab instructor's guidelines when discarding any excess reagents or chemical wastes during or at the conclusion of a laboratory period.

    g. Being careful to discard broken glass only in the container labeled "Broken Glass." *Never* discard broken glass in the regular wastebaskets.

    h. Keeping chemical containers clean and properly labeled.

    i. Cleaning and returning community glassware and equipment.

    j. Putting clothing and other personal effects in their proper place. *Do not* drape them over equipment and work benches. **Place book bags and coats under your lab hood.**

    k. Cleaning your bench top with appropriate spray cleaning solution and straightening up your work area at the end of each laboratory period.

LABORATORY 1

TREATMENT OF SCIENTIFIC DATA

This lab was modified from *Survey of Chemistry 111*, Fourth Edition and used with permission by Jennifer N. Robertson-Honecker.

OBJECTIVES

- Use the rules for determining the proper number of significant figures.
- Make measurements and properly record experimental and calculated values.
- Report derived units with the proper number of significant figures.
- Identify an unknown using intrinsic properties.

INTRODUCTION

Significant Figures: In chemistry and other sciences, there are two types of numbers: *exact* and *measured*.

Exact numbers are numbers with a value that is exactly known. There is no error or uncertainty in the value of an exact number. Examples of exact numbers are numbers obtained by counting individual objects, e.g., 7 eggs in a basket; 5 persons in a room; and numbers set by definition, for example: 12 inches = 1 foot; and 100 cm = 1 m.

Measured numbers are numbers with a value that is not exactly known due to the measuring process. There is always some error or uncertainty in the value of a measured number. The amount of error or uncertainty in a measured number depends on the accuracy of the measuring device used to obtain the number.

The number of digits, *significant figures,* used to represent the measured number reflects the amount of uncertainty in the measured number. The more accurate the measuring device used to obtain the measured number, the less uncertainty in the value of the number and the greater the number of significant figures that are used to represent that measured number. For example, if a student measured a certain length none too carefully with a ruler marked at every centimeter, the student might estimate the length to be 5.7 cm. There are two significant figures in this measured number and due to the

INTRODUCTORY CHEMISTRY LABORATORY MANUAL **1**

LABORATORY 1 TREATMENT OF SCIENTIFIC DATA

measuring process there is some uncertainty inherent in the last digit shown. As a first approximation, the amount of error in the measured number 5.7 cm is ± 0.1 cm (error is assumed to be ± 1 in the last digit shown as a first approximation).

Conversely, if the student measured the same length more carefully with a ruler marked at every tenth of a centimeter, the student might estimate the length to be 5.72 cm. There are three significant figures in this measured number and the amount of uncertainty in the measured number is ± 0.01 cm. This second measurement is a more precise measurement and is correctly written using a greater number of significant figures.

Notice that as the accuracy of the measuring device is increased, the amount of uncertainty is decreased, and the number of significant figures used to represent the number is increased. The bottom line is that the number of significant figures in a measured number gives information about the amount of uncertainty in the measurement.

PART I. DETERMINING THE NUMBER OF SIGNIFICANT FIGURES IN A MEASURED NUMBER

In order to count the number of significant figures (sig figs) in a measured number, do the following. Find the first nonzero digit. That first nonzero digit and all digits to the right (including zeros) of that first nonzero digit are significant if written in a decimal number (a number with a decimal point). Some examples are shown in Table 1.1.

Measured numbers greater than one that end in zeroes must be treated as a special case. For example, the number 60,000 may contain anywhere from one to five significant figures. In such cases, it is best to write the number in standard exponential form. If written as 6×10^4, it has one significant figure; if written as 6.0×10^4, it has two significant figures; 6.00×10^4, three significant figures; 6.000×10^4, four significant figures; and 6.0000×10^4, five significant figures. If a measured number of this type is not written in scientific notation, it is best check with your professor to ask how this case is to be treated.

TABLE 1.1

MEASURED NUMBER	NUMBER OF SIG FIGS	MEASURED NUMBER	NUMBER OF SIG FIGS
5.7 cm	two	6.2 mL	two
5.72 cm	three	6.20 mL	three
5.724 cm	four	0.0078 g	two
5.700 cm	four	7.0078 g	five
17 mg	two	17 chairs	infinite (exact #; no error)
6.02×10^{23} atoms	three	60,000 ng	1 (ambiguous)

PART II. MULTIPLICATION AND DIVISION OF MEASURED NUMBERS

In making calculations, it is important to show the proper number of significant figures in the answer. In multiplication and division, the answer will have the same number of significant figures as the least exactly known of the original numbers.

CONSIDER THE MULTIPLICATION

$$8.0 \text{ cm} \times 5.50 \text{ cm} = 44 \text{ cm}^2$$

In this problem, the number 8.0 has two significant figures, the number 5.50 has three significant figures, and the answer correctly has two significant figures. Notice that the number of significant figures on the answer was limited by the least exactly known number, or in other words by the number with the least number of significant figures.

CONSIDER THE DIVISION

$$\frac{2.00 \text{ g}}{0.30 \text{ mL}} = 0.67 \text{ g/mL}$$

The answer correctly contains two significant figures because one of the original numbers, namely the 0.30 mL, limits the answer to having no more than two significant figures. The same rule applies to multiplication and division of numbers in exponential form.

PART III. ADDITION AND SUBTRACTION OF MEASURED NUMBERS

In addition and subtraction, the absolute error on the answer must be equal to or greater than the absolute error of the original numbers.

CONSIDER THE ADDITION

$$2.4 \text{ g} + 144.06 \text{ g} = ?$$

A mathematics student would most probably express the answer as 146.46 g. In chemistry, this answer would be incorrect because it misrepresents the error. When the answer is expressed as 146.46 g, it indicates an absolute error of ± 0.01 g; a much smaller absolute error than justified by one of the original numbers, 2.4 g (an absolute error of ± 0.1 g).

The expressed final answer must have the same absolute error as the least precise measurement.

In this example, the least precise measurement is 2.4 g and the answer must be rounded at the first decimal place and expressed as 146.5 g.

LABORATORY 1 TREATMENT OF SCIENTIFIC DATA

CONSIDER THE ROUNDING

$$\begin{array}{r} 2.4\text{g} \\ +\ 144.0\,|\,6\text{ g} \\ \hline 146.4\,|\,6\text{ g} \rightarrow 146.5\text{ g} \end{array}$$

Notice that the original number with the least number of decimal places limited the number of decimal places on the answer. The same rule applies to subtraction of measured numbers.

LABORATORY 2

THINK METRIC

OBJECTIVES

- Use measurement to obtain information about objects and substances.
- Measure length, volume, and mass using the metric system.
- Calculate volumes and densities using the metric system.
- Identify an unknown liquid based on its measured density.
- Use significant figures properly in scientific calculations.

INTRODUCTION

This experiment is designed to reacquaint you with the parts of the metric system used by chemists and other scientists. You will be measuring lengths, volumes, and masses. Consult your chemistry textbook for a discussion of the metric prefixes and common units and become familiar with their approximate equivalence in the English system.

The metric system has the advantage of simplicity in converting from one unit of length or mass or volume to another by simply multiplying or dividing by factors of ten, one hundred, one thousand, etc. Also, numerical scales or rules based on the decimal system are much easier to use than ones divided into 1/8's, 1/16's or 1/32's as in the English system.

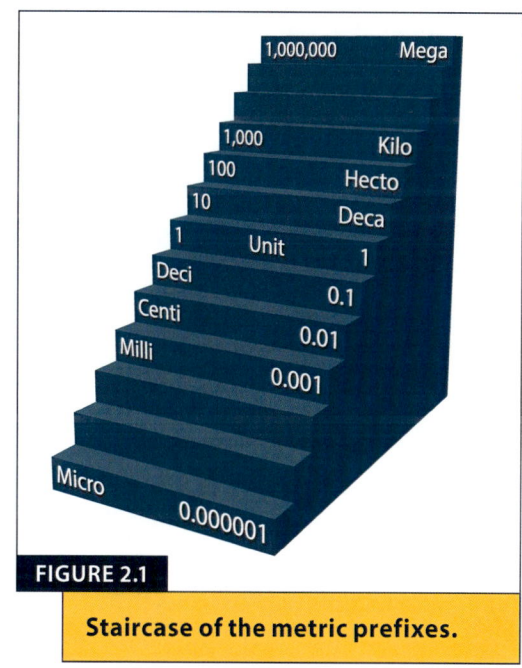

FIGURE 2.1

Staircase of the metric prefixes.

Derived units for properties of materials are also more easily dealt with in metric units. These properties include density, molar volume, or heat capacity, and are all calculated from simple physical measurements.

Consider a simple numerical scale that is divided in units of ten so it can be read in decimal units. Examine a metric ruler and note which units are marked off. Usually the marks for centimeters (cm) are numbered. Each cm is divided into ten smaller units. These smaller units are millimeters (mm) and are equivalent to about 1/25 of an inch. This is the smallest practical spacing for a numerical scale; this applies to thermometer scales and others as well.

Now, if conditions are good, one can usually *estimate* one additional digit between the smallest units marked on the scale. In the illustration below, a centimeter ruler (not to scale) is being used to measure a piece of paper. Careful consideration would give a value of 2.52 cm (or 25.2 mm) for this reading rather than 2.5 or 2.6 cm (or 25 or 26 mm). **In this course, you should always read a scale to get a value in which the last digit is interpolated between the smallest markings.** The last "2" in 2.5 above is estimated between 2.5 and 2.6.

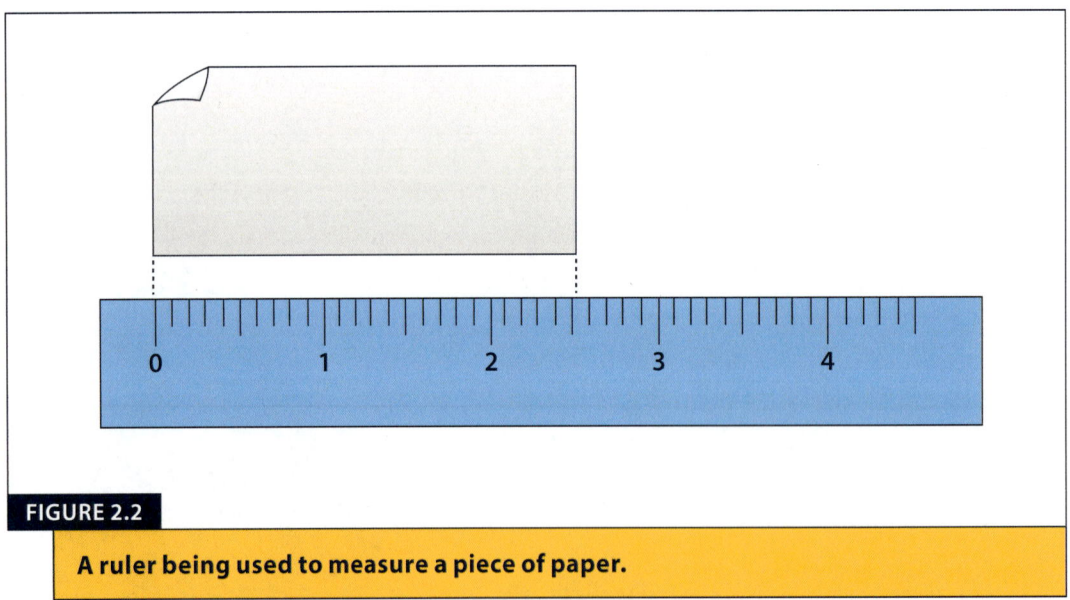

FIGURE 2.2

A ruler being used to measure a piece of paper.

Scientists recognize that the estimated final figure is subject to more uncertainty than the other digits. For example, you might have read the length of the paper as possibly 2.51 cm or 2.54 cm rather than 2.52 cm.

A thermometer scale would present a similar situation. The thermometers that you will be using in this course are calibrated in intervals of 1 °C, and thus temperature readings are typically values estimated to 0.1 °C.

The amount of information and the value of measured numbers depend on their source and the way they are handled. The amount of information in these numbers is expressed using "significant figures." Scientific numbers need to be expressed with just the right amount of information—not too much and not too little.

Density (D) is defined as mass (m) per unit volume (V). It is a physical property of matter that can be quite valuable in characterizing or identifying that substance. For example, we relate regularly to materials that are more or less dense than water, and therefore they sink or float. Pure substances have a very specific density (recall the density of water?), and density can be used to help identify substances. Density is also used to characterize mixtures. An auto mechanic will measure the density of engine coolant to determine the percentage of antifreeze and water in the coolant to be sure the engine is properly protected from freezing.

Density values have units, and in chemistry they are usually grams per cm^3 (g/cm^3) or grams per mL (g/mL). (In the English system, the situation is complex: pounds/ft^3, pounds/gal, pounds/yd^3, and pounds/bushel are just a few that are used.)

This experiment involves determination of densities of several substances for the purpose of identification of unknowns. In each case, measurement leads to identification of the mass and the volume of a sample. Then density is easy to calculate. The experiment requires determination of the volume of a regular object, an irregular object, and a liquid.

PROCEDURE

Record the data from the measurements you make in ink directly on your Report Pages as you go along. Show calculations and answer discussion questions in the spaces provided.

LABORATORY 2 THINK METRIC

PART I

FIGURE 2.3

Measuring the height of a wood block.

WOOD BLOCK DENSITY

1. Obtain a block of wood and record its letter.

2. Carefully measure the length, width, and height of the block to the nearest 0.01 cm. Remember that you *must* interpolate between the two smallest markings on your measuring devise. Report the result in units of centimeters *and* millimeters.

3. Obtain the mass of the same block of wood. You may see some uncertainty in the last digit as the balance reading changes up or down in the hundredths place. Record the best value.

4. Using the formula for density, determine the density of your block of wood and record your conclusion on the board under the appropriate block series.

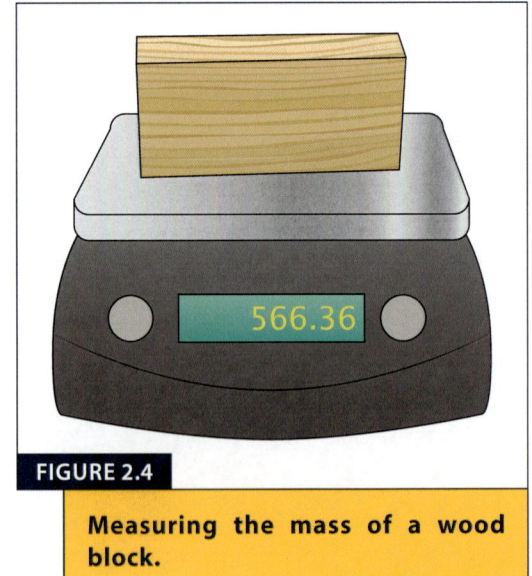

FIGURE 2.4

Measuring the mass of a wood block.

10 INTRODUCTORY CHEMISTRY LABORATORY MANUAL

Note: Density is a characteristic property of a substance and can often be used for identification purposes. Samples of a given kind of wood should have similar densities. These densities differ from the densities of other kinds of wood. Your objective is to compare densities for several samples of wood and to try to identify an unknown sample of wood.

TABLE 2.1

WOOD	DENSITY
Balsa	~0.1
Maple	~0.7
Cherry	~0.8
Oak	~0.8
Ebony	~1.2

PART II

VOLUME OF A CYLINDER USING A RULER

1. Obtain a cylindrical can from the side shelf. Measure and record, in cm, the *inside* height and diameter of the can. Recall that a scale should be read so that you interpolate between the smallest markings.

2. Calculate the volume of the can using the formula:

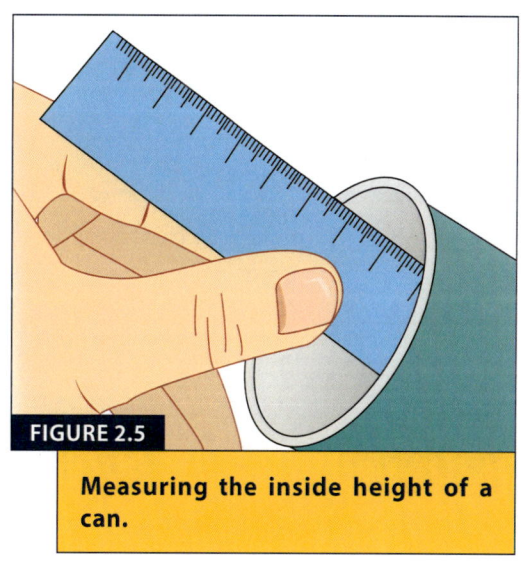

FIGURE 2.5

Measuring the inside height of a can.

$$V = \pi r^2 h \text{ or Volume} = \pi \times (\text{radius})^2 \times \text{height}.$$

VOLUME OF A CYLINDER USING A GRADUATED CYLINDER

1. Fill the can with water to the same height that you measured.

2. Carefully measure the volume of the water in the can by pouring it into the 100-mL graduated cylinder. Note any sources of error that may have occurred as you pour the liquid.

Note: The volume of the can may exceed the volume of the graduated cylinder.

LABORATORY 2 THINK METRIC

PART III

DENSITY OF A LIQUID AND IRREGULAR SHAPED OBJECTS

1. Weigh a clean, dry 10-mL graduated cylinder.
2. Carefully add between 4 and 5 mL of one of the liquids (A, B, C, or D). Read the liquid level at the bottom of the meniscus and be sure to interpolate between the two smallest markings.

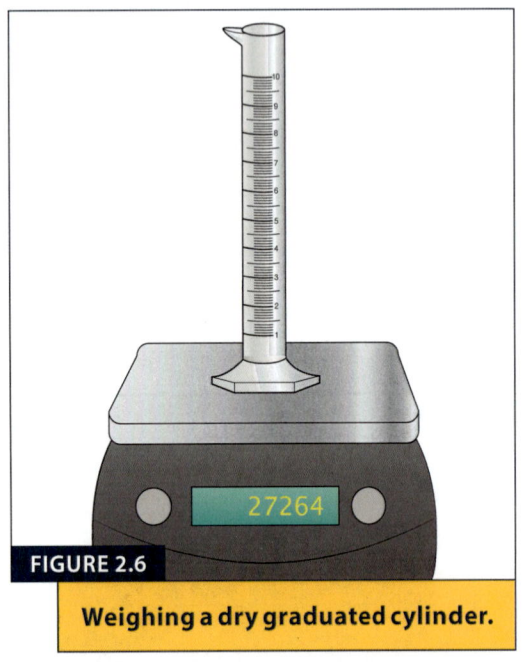
FIGURE 2.6 Weighing a dry graduated cylinder.

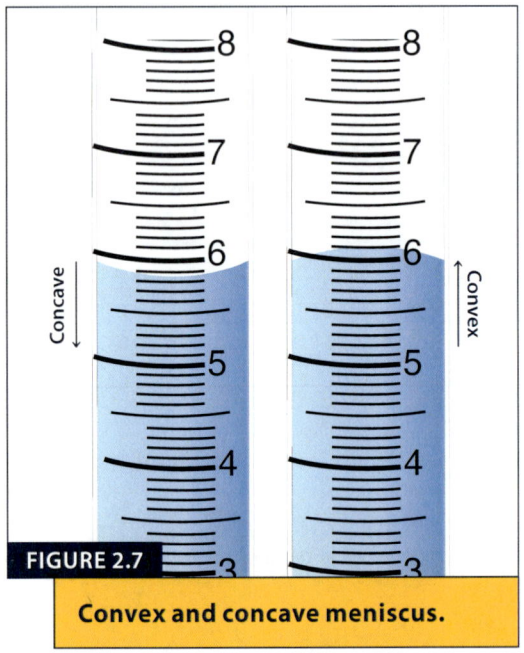
FIGURE 2.7 Convex and concave meniscus.

3. Be sure the outside of the cylinder is dry and then reweigh the cylinder with its contents. The difference between the two masses is the mass of the liquid.

   Discard the liquid into the collection container in the hood.

4. Calculate the density of the liquid. Compare the value obtained with the values below in order to identify the liquid you used.

TABLE 2.2

LIQUID	DENSITY
Acetone	0.79 g/mL
Water	1.00 g/mL
Ethylene glycol	1.11 g/mL
Dichloromethane	1.33 g/mL

Notice how simple measurements can be used to identify an unknown pure substance.

12 INTRODUCTORY CHEMISTRY LABORATORY MANUAL

DENSITY OF IRREGULAR SHAPED SOLID

The determination of the density of an irregular object is based on Archimedes Principle of Displacement. An object that sinks will displace a volume of water equal to the volume of the object. These principles are still fundamentally important in science today, especially forensic science.

1. Place about 5 mL of water in your 10-mL graduated cylinder. Measure this volume to the nearest 0.1 mL.
2. Mass a small amount of irregularly shaped solid on the balance. The amount you mass must not exceed the capacity of the water in the graduated cylinder.
3. Carefully transfer the solid into the graduated cylinder. All the solid **must** be submerged in the water. Tap the sides of the cylinder to release any trapped air bubble. Measure the new volume of liquid in the graduated cylinder.
4. Calculate the density of the material. Compare the experimental density.

FIGURE 2.8

A beaker filled with irregular shaped objects.

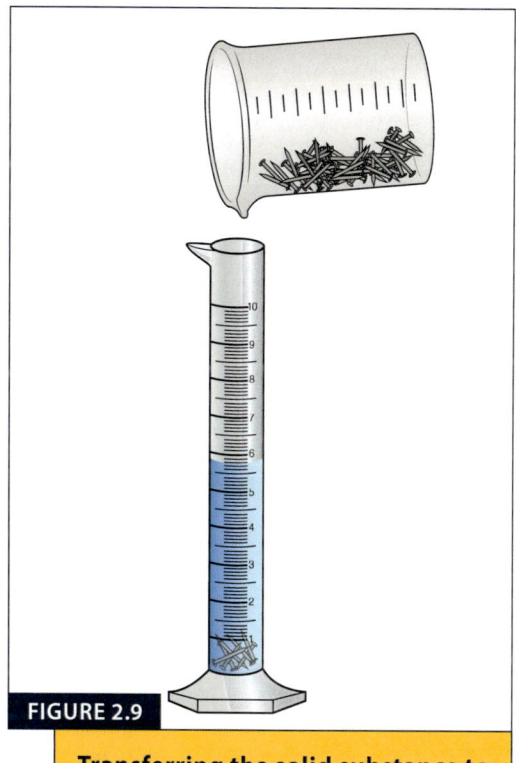

FIGURE 2.9

Transferring the solid substance to a graduated cylinder.

TABLE 2.3

COMMON SUBSTANCE	DENSITY
Aluminum	2.699 g/cm^3
Copper	8.92 g/cm^3
Iron	7.85 g/cm^3
Lead	11.34 g/cm^3
Mercury	13.546 g/cm^3
Tin	7.28 g/cm^3
Zinc	7.14 g/cm^3
Brass (60% Cu, 40% Zn)	8.4 g/cm^3
Glass	2.4–2.8 g/cm^3
Diamond	3.0–3.5 g/cm^3
Rubber	0.9–1.2 g/cm^3
Sugar	1.59 g/cm^3

5. Pour both the unused dry solid and the recovered wet material into the appropriate funnel in the hood.

LABORATORY 3

SUGAR CONTENT OF SOFT DRINKS

OBJECTIVES

- Prepare standard solutions of varying concentration.
- Prepare a calibration curve to relate concentration of a substance to its density.
- Make a scientific decision based on experimental data.

INTRODUCTION

One of the most controversial topics today is whether soda machines should be allowed in schools. Soda and sports drinks offer virtually no nutritional value to students and in many cases, further exacerbate problems such as type II diabetes and childhood obesity. Do you know how many grams of sugar (fructose) are in a 20 ounce bottle of pop?

Sugar contributes more than any other solute to the density of soft drinks. Various soft drinks and juices typically contain two types of sugar: sucrose ($C_{12}H_{22}O_{11}$) and fructose ($C_6H_{12}O_6$). Both sucrose and fructose have very similar aqueous densities, so standard solutions of sucrose can be used as an indicator of the % sugar in various soft drinks.

In this experiment, you will prepare and determine the density of 5, 10, 15, and 20% sugar solutions. A standard graph will be constructed to determine the relationship between the density and the sugar content and be used to calculate the % sugar in your soft drinks.

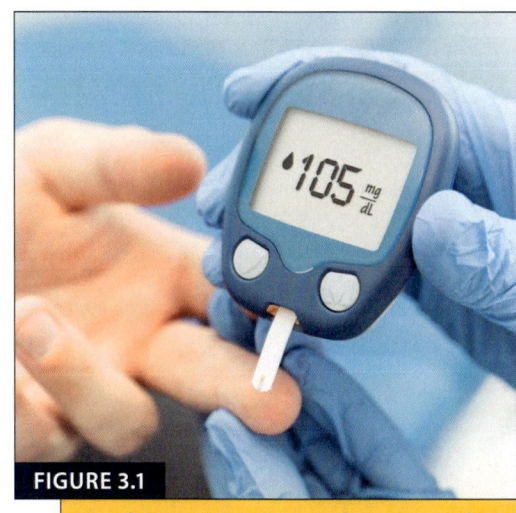

FIGURE 3.1

A glucometer is used to test the sugar levels in a diabetic person's blood.

INTRODUCTORY CHEMISTRY LABORATORY MANUAL **19**

LABORATORY 3   SUGAR CONTENT OF SOFT DRINKS

PROCEDURE

PREPARING YOUR STANDARDS

You will prepare 5 standard solutions (0, 5, 10, 15, and 20% sugar solutions). This procedure describes the preparation of the 5% solution. You will use the same methods for the other solutions, just vary the mass of the sugar used.

1. To prepare a 5% sugar solution, weigh out 5 g of table sugar and record the exact mass in the data table. (This mass of sugar is also equal to the % sugar in your solution. For example, if your mass is 4.77 g, then your prepared % sugar is 4.77%.)

2. In a beaker, combine this table sugar with approximately 75 mL deionized H_2O and stir until **completely** dissolved.

3. Pour the solution into a 100-mL graduated cylinder and add deionized H_2O up to a total volume of 100.0 mL.

4. Transfer the solution into a labeled beaker for later use.

5. Repeat the above steps with 10, 15, and 20 grams of sugar.

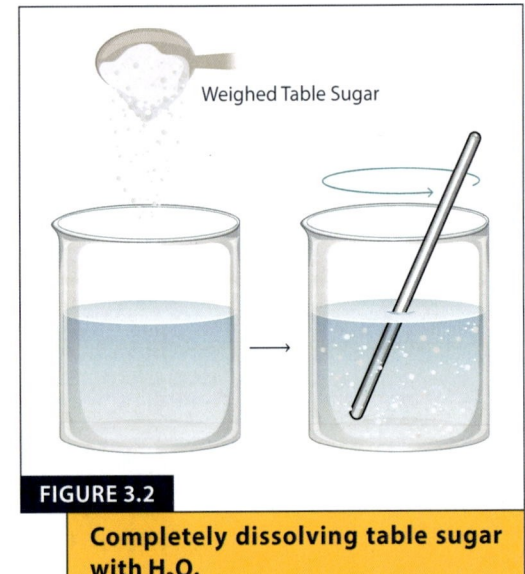

FIGURE 3.2
Completely dissolving table sugar with H_2O.

DETERMINE THE DENSITIES OF THE STANDARD SUGAR SOLUTIONS

1. Using a volumetric pipette, determine the mass of 10.00 mL of sugar solution.

2. Record this mass in the data table for each of your solutions.

3. Calculate the densities of your standards based upon the measured mass of the solution and the dispensed volume.

DETERMINE THE DENSITIES OF THE AVAILABLE BEVERAGES

1. Using the 10.00 mL volumetric pipette, determine the densities of at least two available beverages and one unknown.

2. Record the nutrition information from the beverage label, if applicable.

20 INTRODUCTORY CHEMISTRY LABORATORY MANUAL

CALCULATIONS

1. Calculate the density of each standard solution using the mass of the solution and the volume.

2. Using an available software program, graph density vs. % sugar. Include a linear regression line in your analysis.

3. Using your calibration curve of the standard solutions, determine the % sugar in each of the soft drinks you tested. (Use your linear regression equation to calculate the % sugar in the soft drinks.)

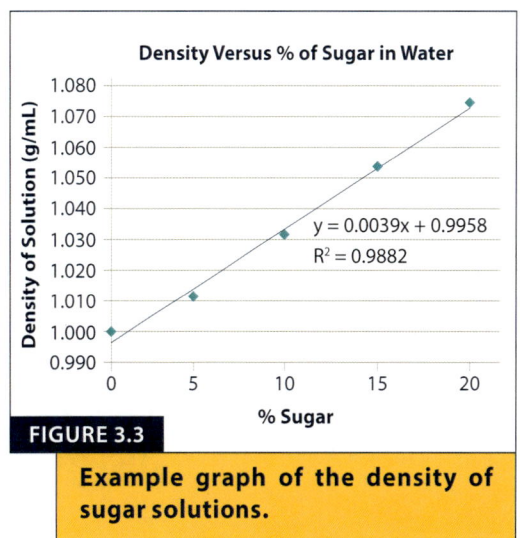

FIGURE 3.3 Example graph of the density of sugar solutions.

LABORATORY 4

THE ENERGY CONTENT OF FOOD

OBJECTIVES

- Relate chemical investigations to everyday experiences.
- Make laboratory measurements and manipulate data.
- Analyze experimental techniques and limitations.
- Determine the energy content of foods.

INTRODUCTION

These days there is a lot of talk about the characteristics of food. Carbohydrates, fats, causes of obesity, diet sodas, and artificial sweeteners are all part of our everyday culture. Consumers often are found reading the food label. Some common entries on food labels are "Calories" and "Calories from fat." The typical recommendation is that a diet should not exceed about 2000 calories per day and should contain less than 65 g of fat. In this experiment, the number of calories involved in burning foodstuffs will be investigated.

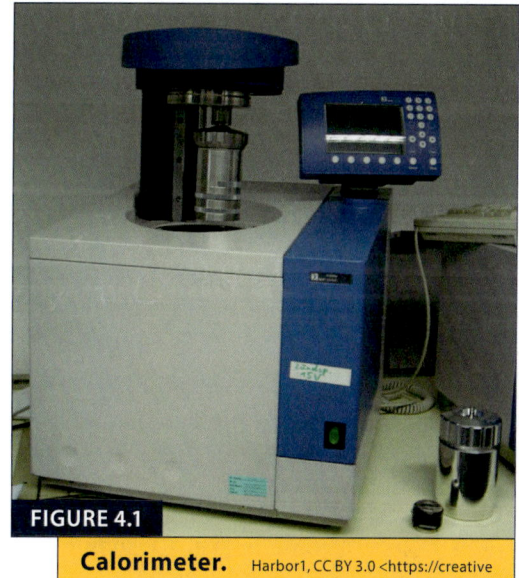

FIGURE 4.1

Calorimeter. Harbor1, CC BY 3.0 <https://creativecommons.org/licenses/by/3.0>, via Wikimedia Commons

The basis for the experiment is the First Law of Thermodynamics—energy is conserved. The experimental method used to measure the heat change associated with a chemical or physical change is called calorimetry. The amount of heat gained or lost by a chemical or physical change is determined from the amount of heat lost or gained by water in a container. A relatively simple container will be used here—a soda can partially filled with water.

INTRODUCTORY CHEMISTRY LABORATORY MANUAL **27**

LABORATORY 4 THE ENERGY CONTENT OF FOOD

The calorimetry experiment involves calculation of the amount of heat absorbed by the water in the calorimeter, and is, therefore, based on the properties of water. Specific heat is the amount of energy necessary to change the temperature of 1 gram of a substance by 1 °C. The specific heat of water is 1.00 calorie per gram per °C, 1.00 cal/g °C. The specific heat of aluminum is considerably less—0.22 cal/g °C, and the specific heat of iron is only half of that. Based on the low mass and low specific heat for aluminum, the heat change for the can will be ignored.

The total amount of heat gained by a substance like the water in the can is calculated from the relationship

$$\text{Heat change} = m \times c \times \Delta T$$

(m is the mass of the substance, c is its specific heat, and ΔT is the temperature change).

By measuring the mass of the water and the temperature change and knowing the specific heat, the total heat change for the water can be determined. *The heat change for the food is the same magnitude but in the opposite direction as the heat change for the water* (The First Law):

$$\text{Heat gained by the water} = m \times c \times \Delta T = -\text{Heat lost by the food.}$$

By following this procedure, the calculated outcome is in calories of heat energy. Conversion to kilocalories should be straightforward (1000 cal = 1 kilocalorie). The nutritional quantity given on food labels is in food "Calories." **One food Calorie is equal to one kilocalorie.**

In the experiment, the heat energy given off by burning food is assumed to be equivalent to the heat energy gained by water in the calorimeter. In the crude setup used here, some heat will be lost to the surrounding air, and some heat will go into the aluminum can, but the specific heat of these is small, and the results are fairly reasonable.

PROCEDURE

1. Obtain a soda can, paperclip, and piece of aluminum foil about 4 cm on a side. Set up a ring stand and clamp.

2. Bend the paperclip so one end points up and the rest serves as a base. Fold some aluminum foil on the base to improve its stability and to allow it to catch the food if it falls. Set the paperclip on the ring stand base.

3. Clamp the soda can and mount the clamp on the ring stand so its bottom is about 2–3 cm above the top of the paperclip.

LABORATORY 4 THE ENERGY CONTENT OF FOOD

4. Obtain a sample of food and poke it onto the paperclip point. This may take some care. Weigh the paperclip assembly with the food.

5. Remove the can from the clamp, add about 100 mL of water, and weigh the can and water. Mount the can back into the clamp. Clamp the temperature probe so it is suspended in the water in the can.

6. Open the *LoggerPro* "Food Calories" icon from the computer desktop.

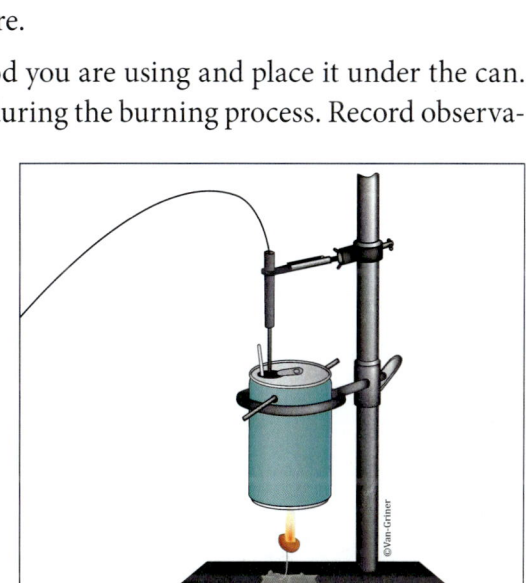

FIGURE 4.2 Food sample on a paperclip.

7. Start data collection by hitting the green arrow. Note the initial temperature.

8. Using the Bunsen burner, ignite the food you are using and place it under the can. Keep the flame centered under the can during the burning process. Record observations. Allow the data to collect and the food to burn until the flame goes out. Wait until the temperature reaches a maximum and remains fairly steady. Note the final temperature and stop the data collection.

9. Reweigh the paperclip assembly to obtain the mass of the food that burned.

10. Repeat your experiment two more times with the same type of food. Empty the soda can and start with fresh water each time.

11. Return the clean paperclip and can to the side shelf, and return the other equipment.

FIGURE 4.3 Set-up of the soda can and food sample experiment.

12. Calculate the heat content of the foodstuff as indicated on the Report Form. Show a sample calculation between the lines of the data table. Post your results on the blackboard for comparison to other results.

13. Record the calorie information from the food package, including serving size in grams.

INTRODUCTORY CHEMISTRY LABORATORY MANUAL 29

Specific heat = $C_{H_2O} = 1.0 \, cal/g°C$

$C_{Aluminum} = 0.22 \, cal/g°C$

heat of combustion = Q

Q = q/mass of food lost while burned

1000 cal = 1 kcal = 1 Cal

LABORATORY 5

SEPARATION OF A MIXTURE

OBJECTIVES

- Learn various separation techniques by separating the components of a mixture.
- Calculate the percent by mass of each component in the mixture.
- Calculate percent recovery.

INTRODUCTION

A pure substance is a single kind of matter with a distinct set of chemical and physical properties. A mixture is a physical combination of two or more pure substances. The pure substances in a mixture retain their physical and chemical properties. The unique physical properties of the substances in a mixture can be utilized to separate the substances into their pure states. Physical separation only involves physical changes in the pure substance, and therefore, does not alter its identity.

In this experiment, you will rely on differences in properties to separate silicon dioxide (SiO_2, sand), sodium chloride (NaCl, table salt), and ammonium chloride (NH_4Cl), from a mixture of these solid compounds. Three commonly-used methods of separation are given below:

SUBLIMATION

Sublimation is the process by which a solid changes from the solid to the gaseous state directly without forming a liquid. *Melting* is a process by which a solid changes to a liquid by heating. In the mixture used in this experiment, one compound, ammonium chloride, sublimes easily, while the other two components do not.

LABORATORY 5 SEPARATION OF A MIXTURE

SOLUBILITY

The extent to which a substance is soluble in a solvent depends upon the chemical structure of both the substance and the solvent. In general, polar compounds, such as sugar and alcohol, and ionic compounds, such as KCl, NaCl, NH_4Cl, and NH_4NO_3, are soluble in polar solvents such as water. Nonpolar substances such as grease, wax, and oil, are soluble in nonpolar solvents such as toluene or kerosene. Extracting (dissolving) a soluble substance out of a mixture with an appropriate solvent is a common separation technique. In this experiment you will use solubility to extract a solid, NaCl, which is soluble in water, from another solid, SiO_2, which is insoluble in water.

DISTILLATION

If two components have very different boiling points, the substance with the lower boiling point will evaporate more rapidly at a given temperature than the substance with a higher boiling point, and so they can be separated on this basis. Solid NaCl with a very high boiling point can be separated from a solution (NaCl—H_2O solution) by simply evaporating the water, which has a much lower boiling point. The solid NaCl will remain in the dish as a dry residue. If the water vapor is condensed to a liquid and is collected, the process is called *distillation*.

The separation techniques of sublimation, extraction, and evaporation usually do not change the chemical composition of a substance, and, therefore, standard separation techniques used by chemists.

PROCEDURE

PART I

1. Obtain a sample of "unknown mixture" and record its identity number or letter on the Report Form.

2. Weigh a clean, dry evaporating dish (or beaker) to the nearest 0.01 g. Record the mass on the Report Form.

3. Measure approximately 2 g of your mixture by carefully adding it directly to the evaporating dish while it is on the balance. The mass of sample does not have to be exactly two grams, but be sure to record all the decimals that the balance reads.

FIGURE 5.1

Weighing an empty evaporating dish.

34 INTRODUCTORY CHEMISTRY LABORATORY MANUAL

PART II

1. Heat the evaporating dish and its contents on an electric hot plate in the fume hood. Ammonium chloride, NH$_4$Cl, will sublime from the mixture and appear as a white smoke.

2. Continue heating until the white smoke is no longer produced. At this point, you can stir the hot mixture with a glass stirring rod to ensure complete sublimation of the ammonium chloride.

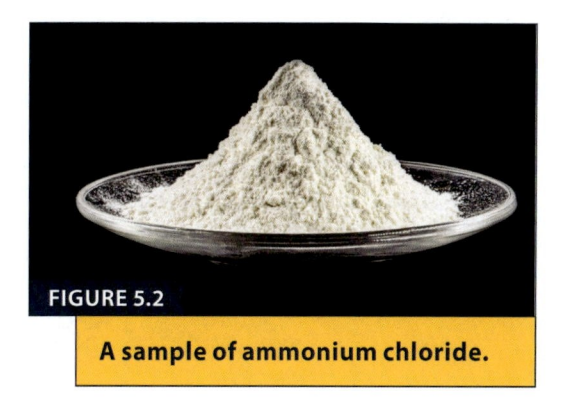

FIGURE 5.2
A sample of ammonium chloride.

3. Allow the evaporating dish and its contents to cool to room temperature and weigh them together. Record this mass on the Report Form. From the differences in masses before and after heating, determine the mass of NH$_4$Cl and calculate its percent by mass:

$$\text{Percent by mass of "Z"} = \frac{\text{mass of Z}}{\text{total mass of the sample}} \times 100.$$

PART III

1. Add about 20 mL of deionized water to the residue in the evaporating dish to dissolve the NaCl. Stir the mixture well for a few minutes with a glass stirring rod to dissolve the NaCl completely.

2. Next, *decant* the water (now a solution of dissolved NaCl) into a second, pre-weighed evaporating dish (or small beaker if there is not room), leaving the undissolved solid (sand, SiO$_2$) behind in the original evaporating dish.

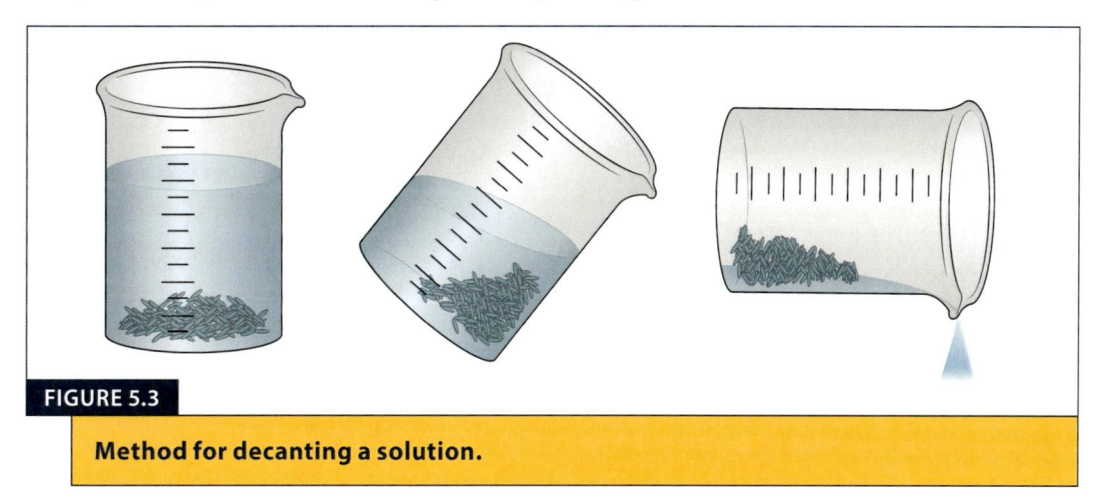

FIGURE 5.3
Method for decanting a solution.

LABORATORY 5 SEPARATION OF A MIXTURE

3. Add about 10 mL of deionized water to the wet sand, stir well with the stirring rod as before, and decant the liquid into the second evaporating dish, which will now contain about 30 mL of water with dissolved NaCl. Set this dish and solution aside for Part IV.

4. Heat the evaporating dish containing the wet sand on the hot plate to dry the sand. Use crucible tongs to quickly lift the dish from the plate when needed, as the last remainder of water can evaporate suddenly and cause the sand to spatter out of the dish. Alternately, you can cover the dish with a watch glass (concave side downward) to prevent any sand from spattering out, but some sand may stick to the watch glass.

5. Allow the evaporating dish and dry sand to cool to room temperature, and if necessary, carefully scrape any sand sticking to the watch glass back into the dish. Weigh the dish and sand and record this mass on the Report Form. Determine the mass of SiO_2 and calculate its percent by mass in the sample.

PART IV

1. Heat the second evaporating dish containing the salt water solution on the hot plate. You should not heat so strongly that the solution splashes out of the container when it is boiling, but it is not necessary to cover the dish with a watch glass, as this will inhibit the evaporation of the water. The goal is to evaporate the water completely, leaving a residue of dry NaCl in the dish. However, when the salt is nearly dry, it has a strong tendency to spatter causing loss of NaCl to occur. There are two ways to prevent this.

   - You can use your crucible tongs to lift the dish off and on the hot plate to control the rate of evaporation of the last of the water, or,
   - You can place a watch glass on the evaporating dish and let any spattering salt stick to the underside of the watch glass.

   Either way, after the salt appears dry and no more spattering is occurring, continue heating the open dish, and the watch glass also if it has any salt adhering to it, for about 10 minutes. This extra heating is done to ensure that the salt is completely dry.

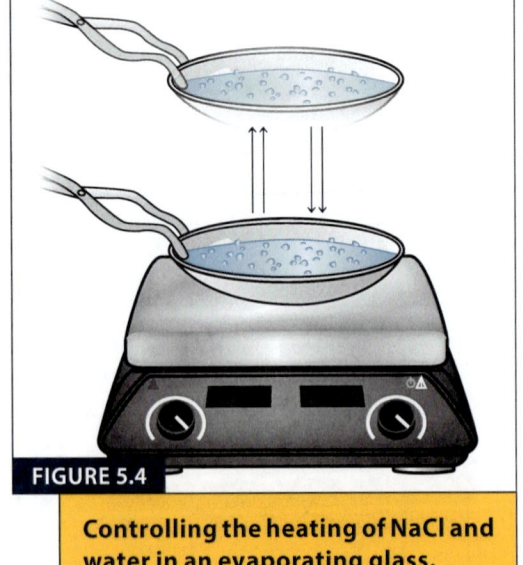

   FIGURE 5.4 Controlling the heating of NaCl and water in an evaporating glass.

2. Allow the evaporating dish and dry NaCl, and the watch glass if it was used, to cool to room temperature. Carefully scrape any salt adhering to the dry watch glass back into the dish.

3. Weigh the dish and salt to the nearest 0.01 g and record this mass on the Report Form. Calculate the mass of NaCl and its percent by mass in the sample.

PART V

1. Add up the masses of NH_4Cl, SiO_2, and NaCl from Parts II, III, and IV, and record this total mass on the Report Form. From this, calculate the total percentage of mixture recovered in this experiment:

$$\text{Percent recovery} = \frac{\text{mass of } NH_4Cl + \text{mass of } SiO_2 + \text{mass of NaCl}}{\text{original mass of the sample}} \times 100.$$

2. Ideally, your percent recovery should be in the neighborhood of 100 percent. If your percentage recovery is less than 100 percent, *or more* than 100 percent, give an *explanation* for the error on the Report Form.

LABORATORY 6
FLAME TESTS AND PROPERTIES OF IONIC COMPOUNDS

This lab was modified from *General Chemistry Prep Workbook*, Third Edition and used with permission by Sumita Singh.

OBJECTIVES

- Investigate the conductivity properties of pure compounds and solutions.
- Classify compounds as covalent or ionic based on their conductivity.
- Identify nonelectrolytes, weak electrolytes, and strong electrolytes based on conductivity behavior.
- Write equations representing dissolution of ionic substances in solution.

INTRODUCTION

Water is the most widely used reaction medium (solvent) in all of chemistry. Water and solutions of substances in water permeate the natural world and play crucial roles in processes in living and nonliving systems. Thus, understanding the role of water in dissolving substances and the nature of substances that are dissolved in water is crucial to chemistry, biology and earth science.

A *solution* is a homogeneous mixture of a dissolved substance (*solute*) in a dissolving medium (*solvent*). In a solution, the solvent separates each individual solute particle down to the molecular level. See the figure showing the process of dissolving sugar such that each individual sugar molecule is separated by the solvent (far right).

INTRODUCTORY CHEMISTRY LABORATORY MANUAL 43

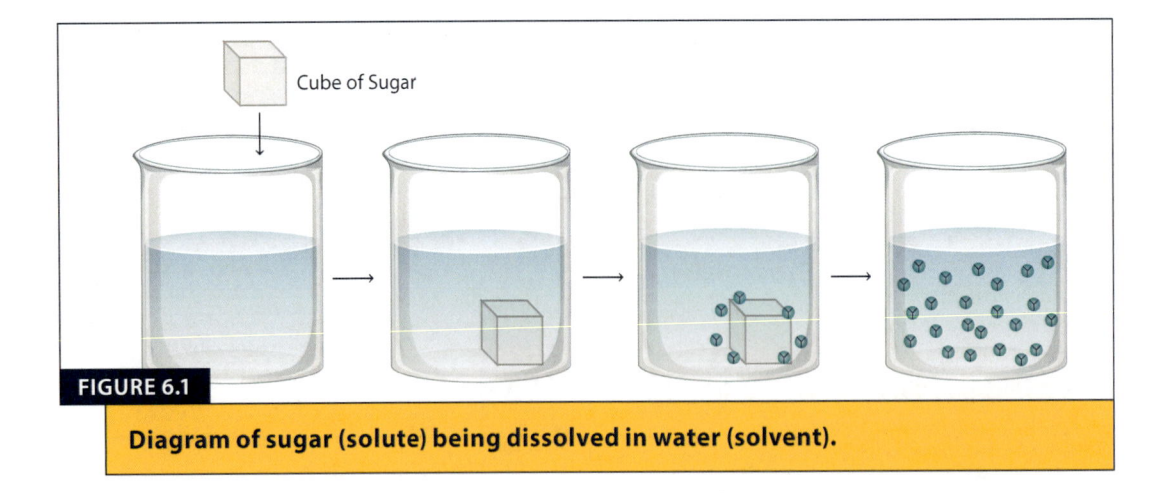

LABORATORY 6 FLAME TESTS AND PROPERTIES OF IONIC COMPOUNDS

FIGURE 6.1

Diagram of sugar (solute) being dissolved in water (solvent).

PART I. FLAME TESTS

Some ions give characteristic colors when they are heated in a Bunsen burner flame. The heat excites electrons from their usual orbitals into higher energy orbitals. When the electrons relax and return to lower energy orbitals, they release energy as light. This effect is also responsible for the colors observed in fireworks. In this part of the experiment, you will identify an unknown cation based on its flame color.

MEET YOUR EQUIPMENT—THE BUNSEN BURNER

- Bunsen burners are often used to heat the contents of a test tube, beaker, crucible, or evaporating dish. They use natural gas to produce a hot flame.

- To light your Bunsen burner, first make sure that the burner is connected by a tube to the gas supply valve on your bench. Adjust the air inlet so that as little air as possible will enter the barrel. Open the gas knob at the bottom of the burner. Open the gas supply valve about ½ way, until you hear hissing gas. Use a match to light the burner at the top of the barrel. You can then open the gas valve the rest of the way and adjust the air inlet. If the flame is orange, more air is needed. If there seems to be a gap between the top of the barrel and the bottom of the flame, close the

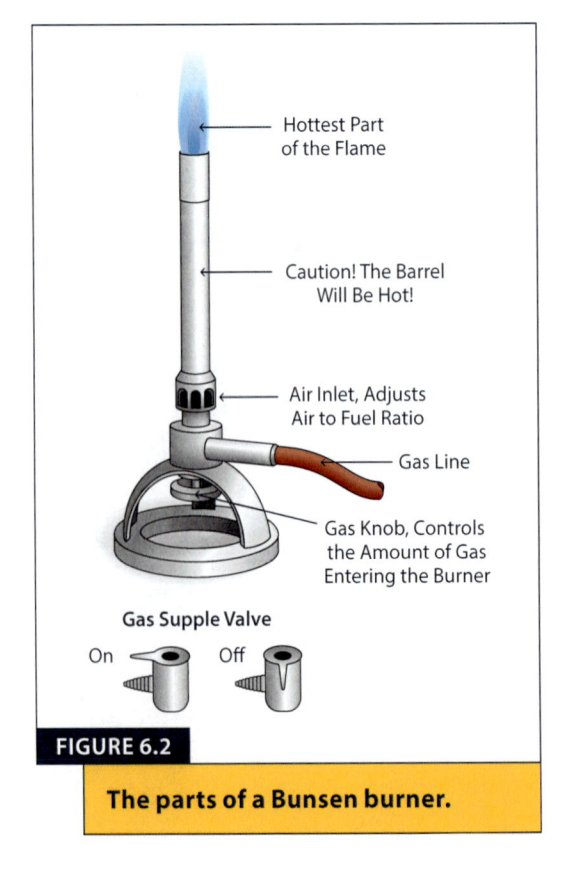

FIGURE 6.2

The parts of a Bunsen burner.

44 INTRODUCTORY CHEMISTRY LABORATORY MANUAL

air inlet a bit. Your goal is a flame that has a purple-blue outer cone and a faint blue inner cone. The hottest part of the flame is the tip of the inner cone, as labeled in the diagram.

> **CAUTION**
>
> If you cannot get the Bunsen burner lit in a few tries, shut off the gas at the valve on the bench, and consult the instructor.
>
> Long hair must be tied back whenever you work with a Bunsen burner.

PART II. CHEMICAL SPECIES IN SOLUTION

When an *ionic compound* dissolves in water, the water pulls the cations and anions away from each other so the cations and anions behave as individual particles, each with its unique properties.

In fact, a nitrate ion in water solution has exactly the same properties whether derived from nitric acid or from potassium nitrate. When an ionic compound dissolves in water, the ions interact only with the water, and an aqueous solution results as indicated by the equation:

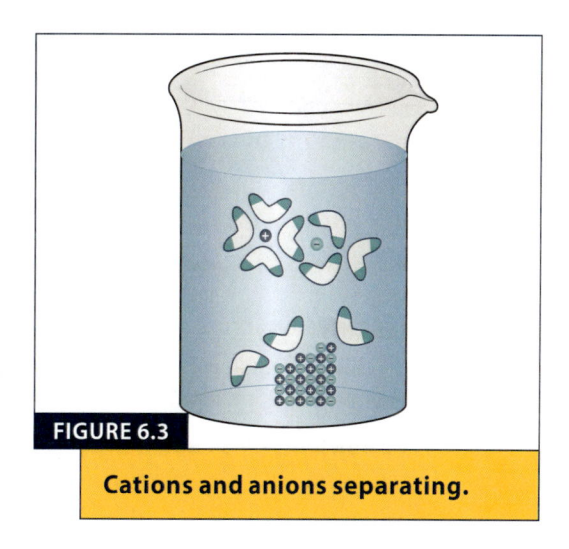

FIGURE 6.3
Cations and anions separating.

$$NaNO_3(s) \rightarrow Na^+(aq) + NO_3^-(aq).$$

The (*aq*) symbol emphasizes that the ions are dissolved species and have properties different from the original solid (*s*). Each ion is its own individual chemical species, unlike the reactant where the ions were ionically bonded together.

When *covalent compounds* dissolve in water, each individual molecule separates from all other solute molecules and behaves as individual unit. Usually they dissolve in water without undergoing any reaction. For example, pure liquid ethanol dissolves in water to form solvated ethanol molecules:

$$CH_3CH_2OH(l) \rightarrow CH_3CH_2OH(aq).$$

In some cases *covalent solute molecules* undergo *ionization* when they dissolve. Certain compounds such as the binary hydrogen halides undergo *complete ionization* in water as shown by the equation for hydrogen chloride dissolving in water:

$$HCl(g) \rightarrow H^+(aq) + Cl^-(aq).$$

Other compounds like acetic acid dissolve and undergo only partial ionization:

$$HC_2H_3O_2(aq) \rightleftarrows H^+(aq) + C_2H_3O_2^-(aq).$$

LABORATORY 6 FLAME TESTS AND PROPERTIES OF IONIC COMPOUNDS

The symbol ⇌ implies a dynamic equilibrium between reactants and products. All three species, $HC_2H_3O_2$, H^+, and $C_2H_3O_2^-$, are present in the mixture. In those cases where only a few ions form from a molecular species such as an acid, the system is represented in an equation by the molecular formula of the major species in solution.

ELECTROLYTES

Solutions that are very good conductors of electricity are called *electrolytes*. Electrolytes conduct because of the presence of ions. Those substances that are *completely ionized* in aqueous solution are called *strong electrolytes,* and those that only *partially ionize* in aqueous solution are called *weak electrolytes*. Those substances that produce *no ions* in solution are called *nonelectrolytes,* and they conduct electricity very poorly.

Solutions of electrolytes conduct electrical current due to movement of their ions as the ions are acted on by an applied electrical potential. For a given applied potential, the amount of current depends primarily on a) the concentration of ions and b) the conductance of a particular ion (some ions are more mobile in the solution than others). The concentration of ions in solution can actually be determined by measuring conductivity.

The concentration of ions in solution will depend on a) the amount of solute added to a solvent, and b) the amount of ionization that occurs when a molecular compound dissociates in solution. So, *conductivity may be low because only a small amount of solute is added even though the substance completely ionizes in solution.*

In summary, a) if ionic compounds dissolve, they usually do so as separate cations and anions and they are strong electrolytes, b) if molecular compounds dissolve as molecules, they behave as nonelectrolytes, and c) some molecular compounds undergo complete or partial ionization when they dissolve.

PROCEDURE

PART I. FLAME TESTS

1. Obtain about 10 mL of 1.0 M HCl in a test tube and 1 mL of each of the other solutions in test tubes.

2. Clean the nichrome wire by dipping it into the HCl and then holding it in the flame of a lit Bunsen burner. If you see a bright orange color, a sodium impurity is present, and the wire should be further cleaned with HCl and heat.

FIGURE 6.4

Process of sterilizing wire with a bunsen burner.

LABORATORY 6 FLAME TESTS AND PROPERTIES OF IONIC COMPOUNDS

3. You will test solutions of KCl, CaCl$_2$, LiCl, BaCl$_2$, and NaCl. For each test, put 2–3 drops of the sample near the edge of the watch glass. Light the Bunsen burner, and open the air intake valve. Hold the end of the nichrome wire in the hottest part of the flame until it glows brightly.

4. Quickly bring the watch glass up to the air intake and touch the hot nichrome wire to the sample. Some of the sample will evaporate and will be pulled into the burner.

5. Observe the characteristic color of the flame with your goggled eyes. Record your observations.

6. Rinse the watch glass and eye dropper thoroughly between samples. Clean the nichrome wire as in Step 2.

TESTING AN UNKNOWN

1. Obtain an unknown from the instructor. The unknown will contain only one of the cations you tested.

2. Test the unknown and record your hypothesis for the identity of the cation. It may be helpful to repeat some of the flame tests again if you are having trouble choosing between two similar colors.

PART II. CONDUCTIVITY

1. Open the Logger*Pro* icon labeled "Conductivity" on the Desktop. Be sure the switch on the control unit is set at 20,000 µS. The unit of measurement for conductivity is Seimens/cm, or microSiemens/cm (µS/cm). It is effectively the inverse of electrical resistance. As the conductivity of the solution increases, its resistivity decreases, and more electrical current will flow.

2. Pour about 5 mL of sample solution (0.1 M, 0.01 M, 0.001 M, 0.0001 M solutions of NaCl and 0.05 M solutions of NaCl, CaCl$_2$, AlCl$_3$) into a well-rinsed, large test tube. Immerse the probe in the solution. (**Note:** You are testing each sample individually, *not* mixed together.) The probe is fragile—*handle it with care*. Record the resulting conductivity including the standard conductivity unit of µSiemens/cm. There will be some fluctuation in the reading. Just record the first two or three significant figures as the best value for conductance.

FIGURE 6.5
Set-up of a probe measuring conductivity.

3. When finished with a measurement, rinse the probe thoroughly with deionized water and gently blot dry on a few paper towels.

4. Prepare a sample of sugar water and dilute ethanol by mixing a small amount of pure sample with deionized water in a beaker. Measure and record the conductivities for each of these samples.

LABORATORY 7

LAW OF DEFINITE COMPOSITION

OBJECTIVES

- Use crucible techniques to perform a high temperature chemical reaction.
- Evaluate the gain in mass when a metal combines with oxygen.
- Calculate percent composition based on experimentally determined masses of components.
- Evaluate the concept of the Law of Definite Composition.

INTRODUCTION

Pure substances fit into one of two categories: they are either elements or compounds. Elements are the fundamental substances, about 118 in number, that are listed in the Periodic Table. They differ from each other by the numbers of subatomic particles (protons, electrons, and neutrons) they contain. All atoms of a given element have the same number of protons in the nucleus. The atomic number of the atom is defined as the number of protons in the nucleus. Elements can combine with one another in a chemical change or chemical reaction to form a compound. A compound, as a pure substance, will have characteristics different from the elements in its makeup. A compound has a definite composition of its constituent elements.

In this lab experiment, the compound magnesium oxide will be prepared from its elements. The chemist's "short-hand" for representing this reaction is the following equation:

$$2Mg + O_2 \rightarrow 2MgO.$$

It uses the symbols for the elements magnesium and oxygen, and the appropriate subscripts and coefficients to account for all of the atoms reacting (i.e., to "balance" the equation). The above equation indicates that two atoms of magnesium react with one molecule of oxygen, O_2 (a molecule of oxygen contains two atoms of oxygen), to form two units of MgO.

LABORATORY 7 LAW OF DEFINITE COMPOSITION

A weighed amount of magnesium will be heated in order to react it with air. Air is a mixture of about 20% oxygen and 80% nitrogen, along with traces of the inert gases, water, and carbon dioxide. The magnesium will have the opportunity of combining with as much oxygen as it can. But only a limited amount of oxygen reacts and, like all compounds, magnesium oxide will form its own unique definite composition. If two students use different amounts of Mg, they will obtain different masses of MgO. Will the two samples have the same composition, or will the % Magnesium increase with sample size?

A common way to express the relative value of quantities is to report percentages. The composition of MgO will be expressed as the percent magnesium in the compound and the percent oxygen in the compound. Percentage can usually be expressed using the following relationship:

[handwritten: H₂O mass of H / mass of H₂O × 100%]

$$\text{Percent} = \frac{\text{part}}{\text{whole}} \times 100\%$$

In determining the percent magnesium in magnesium oxide, the Part of interest is the mass of magnesium. The Whole is the mass of magnesium oxide.

PROCEDURE

CAUTION	The iron ring and crucible will become extremely **hot!** Wait until they cool before touching with your hands!

1. Obtain a *dry* crucible, cover, clay triangle, Bunsen burner, and iron ring from the side shelf. Make sure the crucible has no cracks.

2. Add sufficient sand to cover the bottom of the crucible to a depth of about 0.5 cm (about 1 scoop).

3. Weigh the crucible and sand to the nearest 0.01 g and record the mass directly on your Report Form. **Note:** Do not include the mass of the cover as these commonly fall off and break.

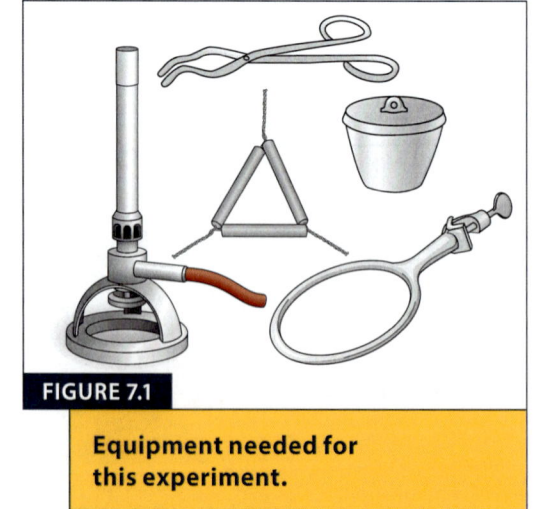

FIGURE 7.1 Equipment needed for this experiment.

4. Add Mg pieces to the crucible until the balance indicates that between 0.4 g and 0.8 g of Mg turnings have been added. Record the exact mass.

5. Set up a ring stand, iron ring, the clay triangle, and crucible near the back of your hood. Open the hood sash all of the way. **Note:** Do not try to readjust the iron ring with the crucible on it. The crucible is likely to fall and break.

LABORATORY 7 LAW OF DEFINITE COMPOSITION

6. Light the burner away from the ring stand and adjust the flame properly. Then move it under the crucible.

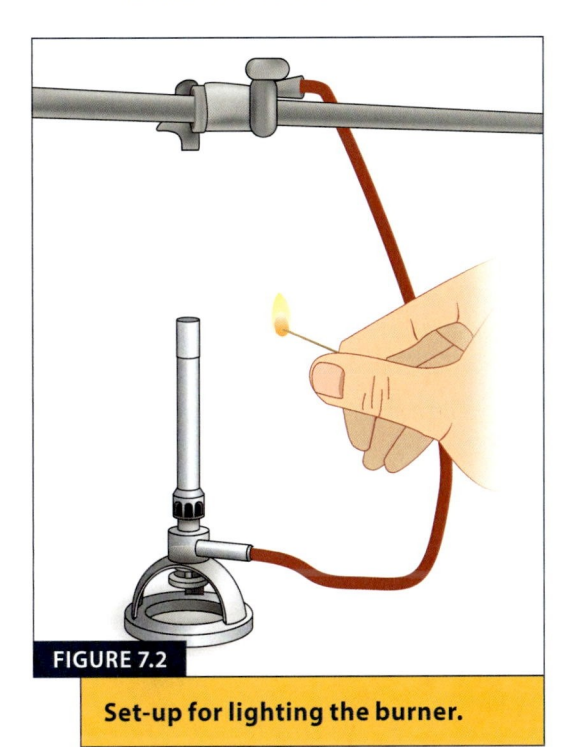

FIGURE 7.2
Set-up for lighting the burner.

FIGURE 7.3
A hot and cool Bunsen burner flame.

7. Heat the **partially covered** crucible gently for about 5 minutes, and then increase the gas flow to continue heating more intensely for 10 minutes longer. The crucible should glow red on the bottom when heated adequately. The magnesium will react with air that enters the crucible. If the sample inflames, use the tongs to close the cover for 10–15 seconds. Then set the cover ajar again to let small amounts of oxygen into the crucible. When heated sufficiently, the magnesium sample will no longer flair up and glow brightly when the cover is lifted slightly.

8. After allowing the crucible to cool on the ring stand for a few minutes, use the tongs to carefully move the cover and then the crucible to the insulation pad. Do not place the hot container directly on the bench top.

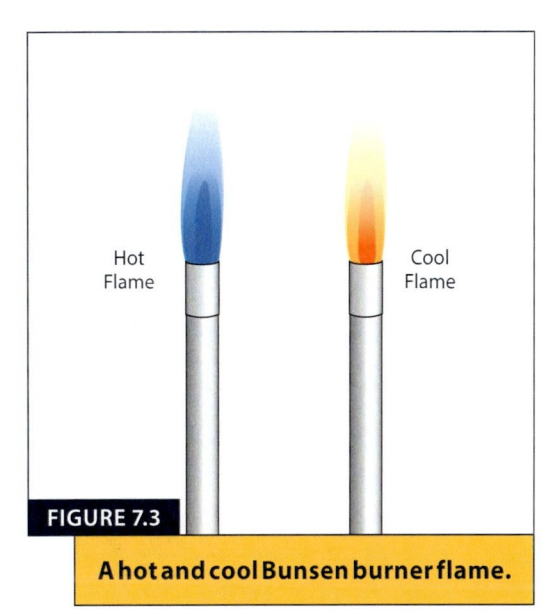

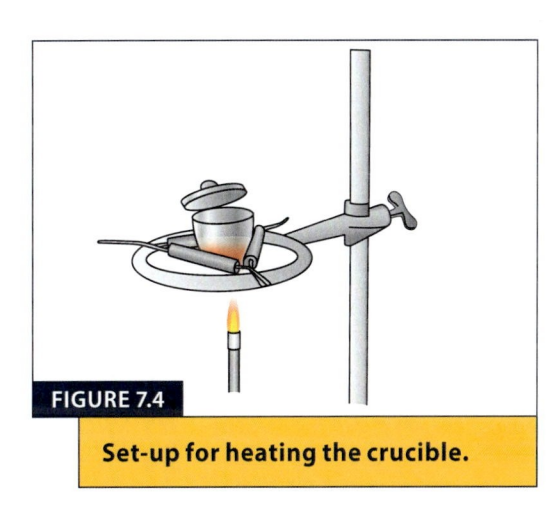

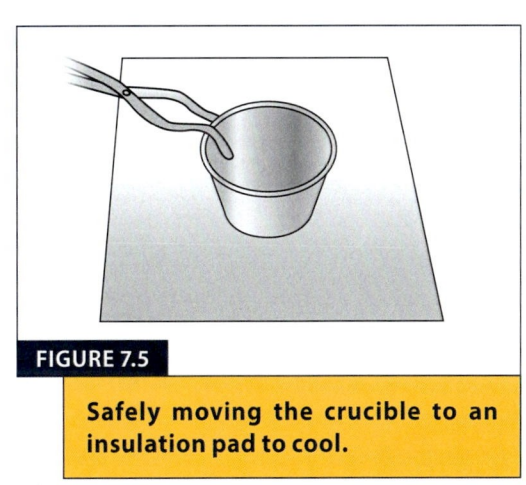

FIGURE 7.4
Set-up for heating the crucible.

FIGURE 7.5
Safely moving the crucible to an insulation pad to cool.

INTRODUCTORY CHEMISTRY LABORATORY MANUAL 55

LABORATORY 7 LAW OF DEFINITE COMPOSITION

9. Allow the crucible to cool to nearly room temperature and then weigh it. Record this mass on the Report Form in the line labeled "Mass of crucible, sand and MgO after first heating."

10. Once again, heat the crucible intensely for 5 more minutes, let it cool, and weigh it. Record its mass. If these two values for mass differ by more than 0.05 g, heat, cool and reweigh the sample again. The objective is to bring the sample to **constant mass** following successive heating, cooling, and weighing cycles. Consult with your instructor if you have difficulty achieving constant mass.

11. Record your mass of Mg used and your mass of MgO obtained on the board. Copy other students data onto your own Report Form.

12. When you are certain you have completed your work, pour the sand and MgO into the waste container provided and wipe with a paper towel. Do *not* wash the crucible with water.

13. Return all equipment to its proper location and clean up your area.

LABORATORY 8

THE PREPARATION OF ALUM FROM SCRAP ALUMINUM

OBJECTIVES

- Use laboratory synthesis methods to prepare a common household chemical.
- Determine the yield and percent yield in the preparation of a chemical compound.
- Relate composition of a compound to its formula.

INTRODUCTION

Chemical compounds composed of a metallic element and one or more nonmetallic elements are usually classified as salts—consisting of a positively charged cation, like Na^+, and negatively charged anions, like Cl^-. Epsom salts ($MgSO_4$), deicing salt ($CaCl_2$), and baking soda ($NaHCO_3$) all fit in this category. Salts are often prepared by crystallizing the salt from an aqueous solution of the appropriate ions. Salt deposits in nature occur when water evaporates from a body of salt water.

In this experiment, a salt compound (alum) will be produced starting with one of the components (aluminum) in elemental form. Aluminum metal will be reacted with potassium hydroxide, KOH, to produce an aqueous solution of $KAl(OH)_4$. Elemental hydrogen gas is produced as a byproduct:

$$2Al + 2KOH + 6H_2O \rightarrow 2KAl(OH)_4 + 3H_2.$$

Addition of sulfuric acid, H_2SO_4, neutralizes the basic $KAl(OH)_4$ to produce a solution of alum, $KAl(SO_4)_2$. When the alum crystallizes from solution, the hydrated salt $KAl(SO_4)_2 \cdot 12\ H_2O$ forms. A "hydrate" contains water molecules trapped in the lattice of the ions in the solid.

The objective in this experiment is to obtain a high yield of alum from the starting aluminum metal. For each 1 gram of aluminum that reacts, the *theoretical yield* of alum is 17.6 grams. Use this conversion factor (17.6 g alum/1 g Al) to predict the theoretical

INTRODUCTORY CHEMISTRY LABORATORY MANUAL **59**

LABORATORY 8 THE PREPARATION OF ALUM FROM SCRAP ALUMINUM

yield of alum based on the mass of aluminum that you used. Finally, the percent yield is defined as the mass of alum actually obtained divided by the theoretical yield, multiplied by 100%:

$$\% \text{ yield} = \frac{\text{actual yield}}{\text{theoretical yield}} \times 100\% = \frac{\text{mass of alum obtained}}{\text{mass of alum possible}} \times 100\%.$$

Your yield should be less than 100% because some alum remains dissolved in the water solution and some is lost in handling.

Alum is used today for hardening photographic film, preparing pickles, creating and fixing dyes (mordant), and making "lakes" (insoluble dyes used in make-up and food when the color absolutely must not run or bleed). Alum is also a component of solutions used to soak sore joints and to reduce swelling. Shaving styptic (used to stop bleeding from nicks and cuts) is usually potassium alum, although aluminum sulfate is sometimes used.

FIGURE 8.1

A block of alum aftershave. Jiteshn, CC BY-SA 3.0 <https://creativecommons.org/licenses/by-sa/3.0>, via Wikimedia Commons

PROCEDURE

PREPARING THE SOLUTION

1. Obtain between 0.80 and 0.90 g of aluminum; this is roughly a 5 cm square piece. Record the exact mass.

2. Cut the aluminum into several small pieces and place them in a clean 250-mL beaker. Add 25 mL of water and 25 mL of 3 M KOH solution.

3. Stir the mixture, and set the beaker on a hot plate to warm slightly. The aluminum should dissolve and give off hydrogen gas. The aluminum should all dissolve in about 10 minutes. There may be a slight suspension of impurities but they will be filtered out later.

4. Chill the beaker in an ice-water bath (prepared by filling a beaker half full of ice and adding tap water up to the level of the ice).

FIGURE 8.2

Heating a beaker of foil on a hot plate and transferring it to an ice bath.

60 INTRODUCTORY CHEMISTRY LABORATORY MANUAL

LABORATORY 8 THE PREPARATION OF ALUM FROM SCRAP ALUMINUM

5. Remove the beaker from the ice and add 20 mL of 9 M H_2SO_4 *carefully* and *slowly* with stirring. **Note:** If alum precipitates after the addition of sulfuric acid, you will need to dissolve the alum by reheating the solution; otherwise, you will filter out your product and clog the long stem funnel.

CAUTION	Sulfuric acid is strong enough to cause serious burns. Wear gloves and dispose of them when you are through handling the acid. Clean up all glassware used in handling the sulfuric acid.

6. Do not re-cool or chill after the addition of the acid. The reaction mixture should be reasonably clear at this point. If any large lumps of aluminum hydroxide are visible, warm the mixture and stir until the gel-like hydroxide dissolves. A slight amount of suspended impurities remaining may be filtered out in the next step; do not cool in the ice bath if you reheat.

The chemical reactions have now been completed and the remainder of the procedure will involve "cleaning up" the solution, crystallizing and collecting the product, and drying and weighing it.

"CLEANING UP" THE SOLUTION

1. Set up a long stemmed funnel in an iron ring.

2. Place a small mat of wet glass wool down in the cone of the funnel and filter your mixture. (Your mixture should be relatively warm from the previous step to prevent crystallization in the funnel.) Catch the filtrate in a clean 150-mL beaker.

3. Cool this solution by placing the beaker in an ice bath, being careful that it is not going to tip over as the ice melts. Stir your solution occasionally to help heat transfer. Wash all your dirty glassware, discarding the glass wool in the trash can.

FIGURE 8.3
Filter funnel setup.

4. After crystals start to appear, continue cooling until the temperature of the mixture has dropped below 5 °C.

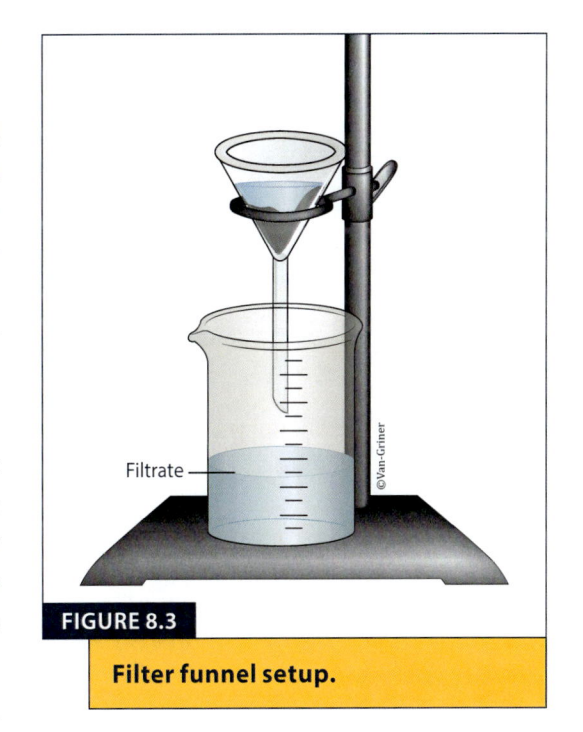

5. Set up the Buchner funnel with a piece of dampened filter paper. Use a ring stand with an iron ring to secure the flask.

INTRODUCTORY CHEMISTRY LABORATORY MANUAL 61

LABORATORY 8 THE PREPARATION OF ALUM FROM SCRAP ALUMINUM

FIGURE 8.4

Set-up of the Buchner funnel with a ring stand.

6. With rapid stirring to keep the crystals from settling to the bottom of the beaker, pour the mixture onto the funnel to collect the solid. Allow the suction to pull off all the liquid. Release the suction by turning off the vacuum and breaking the seal. Rinse the beaker with a quick squirt of deionized water to flush some of the residual alum from the beaker into the filter. Apply vacuum to pull off the water.

7. Rinse the crystals with acetone to remove all the excess water. To do this, stop the suction and pour 10 mL of acetone over the crystals. Let this stand for a few seconds, then reestablish the suction and pull off all the acetone. Repeat with a second 10 mL portion of acetone and this time pull air through the crystals for several minutes to evaporate the acetone.

CAUTION	Acetone is flammable!

8. Weigh your large watch glass and record its mass.

9. Use your white spoon to transfer the dry crystals to the watch glass. Recover as much material as reasonable, but do not try to recover every speck and do not contaminate the product with scrapings of filter paper.

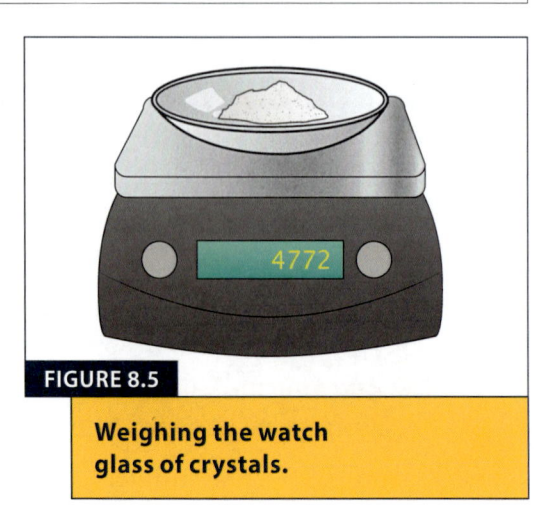

FIGURE 8.5

Weighing the watch glass of crystals.

LABORATORY 8 THE PREPARATION OF ALUM FROM SCRAP ALUMINUM

10. Clean up the funnel assembly by washing the parts in tap water and rinsing with deionized water. Return it immediately so someone else can use it.

11. Allow your alum to dry while carrying out the dyeing experiment discussed below.

12. Determine the total mass of the watch glass and crystals, and then calculate the yield of product.

13. Show the product to the instructor for his/her evaluation and have your report initialed.

CLEAN-UP AND WASTE DISPOSAL

The alum should be placed in the collection bottle.

ALUM AS A MORDANT FOR DYEING (WORK INDIVIDUALLY)

Mordants are additives used in many dyeing processes. In some cases, a mordant serves as the link to bind the dye molecules more tightly to the fabric molecules (L., mordere, to bite). In others, the mordant changes the color of the dye by changing its molecular structure. The use of alum as a mordant with the dye alizarin will be investigated here. Alizarin was used to dye the uniforms of the British "Redcoats" in the eighteenth century.

1. Obtain two pieces of wool. (Use crucible tongs throughout the dyeing process.)

    Note: There will be two alizarin baths—one for wool that was **not treated** with alum and a second one for wool **treated** with alum. This will prevent the one alizarin bath from being contaminated with alum.

2. Immerse one piece of wool in a hot alizarin dye bath ("Not Treated") for two minutes, remove it, rinse it with water, and blot it on a paper towel.

3. Immerse the other piece of wool in the alum bath for 2 minutes. The alum bath has already been prepared for you.

4. After 2 minutes, rinse it with water and then immerse it for 2 minutes in the hot alizarin solution meant for alum-treated wool ("Treated").

5. Remove the wool. Then rinse and dry it.

6. Attach and label both dyed samples on your Report Form.

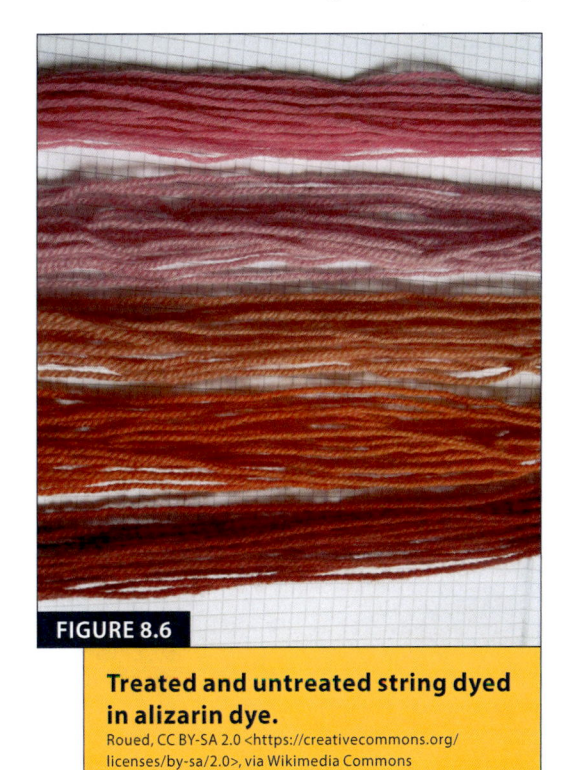

FIGURE 8.6

Treated and untreated string dyed in alizarin dye.
Roued, CC BY-SA 2.0 <https://creativecommons.org/licenses/by-sa/2.0>, via Wikimedia Commons

LABORATORY 9

SAPONIFICATION OF VEGETABLE OIL AND SOAP PROPERTIES

OBJECTIVES

- Learn the method used to prepare soap.
- Relate organic functional groups to chemical and physical characteristics of a substance.
- Transform an oil into a water-miscible substance by functional group transformation.
- Evaluate the ability of soap and detergent to emulsify oil.
- Evaluate the effect of hard water ions on soap and detergent molecules.

INTRODUCTION

PART I. SYNTHESIS OF SOAP

Soap is the most recognized cleaning agent in the world. It was prepared prior to 2000 BC and has been the mainstay cleaning agent until the advent of specially designed chemical compounds for use in modern cleaning agents in the 20th century. For many years, soap was made from animal fat and lye (NaOH).

A soap molecule consists of a long, nonpolar hydrocarbon chain with an ionic group on one end. Notice the sodium cation, Na^+, and the fatty acid $-CO_2^-$ anion.

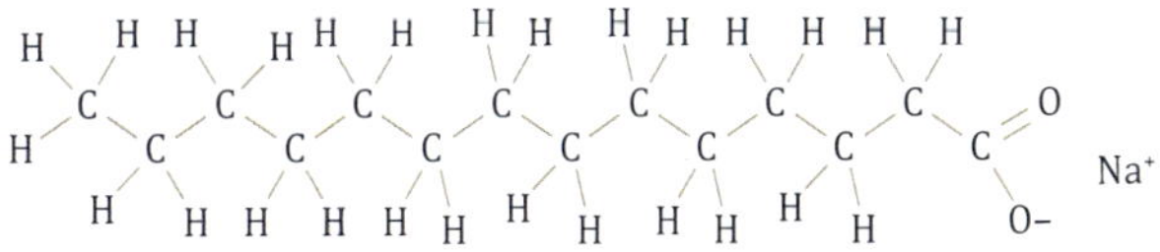

FIGURE 9.1

Molecule chain of soap.

INTRODUCTORY CHEMISTRY LABORATORY MANUAL **67**

The substance can also be described as a carboxylate anion or fatty acid anion, containing a long hydrocarbon tail.

Vegetable oils and fats are chemical compounds containing the combination of three long chain fatty acid groups (the left part of the structure below) linked by ester functional groups (circled) to a glycerol fragment (the right end of the structure). Note that compared to the full structure given for the soap above, this shorthand notation indicates a carbon atom with appropriate number of hydrogen atoms at each line junction. Thus, the top fatty acid chain has 16 carbon atoms and 31 hydrogen atoms in the chain.

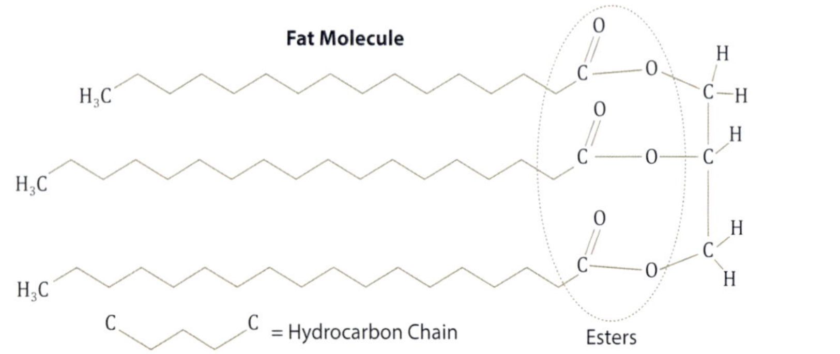

FIGURE 9.2 A fat molecule chain.

Soap is prepared by breaking down the esters of a fat or oil (circled in the structure) into three fatty acid anions and glycerol. The chemical reaction is called **saponification.**

The fatty acid anions in soap are special molecules which contain an electrostatically charged end, but the rest of the molecule is nonpolar. It is the opposing characters (nonpolar on one end and ionic on the other) which makes soap useful as a cleaning material. The nonpolar end of the soap molecule will mix with other nonpolar materials, such as dirty oil and grease. The anion end, being electrically charged, mixes with very polar water. In this way, the oily, greasy dirt is pulled into water by the nonpolar end of the soap molecule. Thus, the oily or greasy dirt, which would not mix with pure water, can be carried away by the soapy water. This is what makes soapy water effective at washing greasy pots and pans.

PART II. SOAPS, DETERGENTS, AND EMULSIFICATION

One of the most common everyday applications of solubility involves cleaning in the home—clothes, dishes, woodwork, windows, and skin. Based on the principle that "like dissolve like," we expect that ionic and polar covalent "dirt" can be washed away with water, a strongly polar covalent solvent which is readily available. However, much of the "dirt" encountered is oily or greasy (nonpolar). Normally, we don't expect nonpolar substances to dissolve in water. We cannot wash away grease and oily dirt with just

LABORATORY 9 SAPONIFICATION OF VEGETABLE OIL AND SOAP PROPERTIES

water. Soap or detergent (surfactants) emulsifies (mixes) the oil and the water. One part of the surfactant molecule has an affinity for non-polar substances (it is *hydrophobic*) while the other end has an affinity for strongly polar water (it is *hydrophilic*).

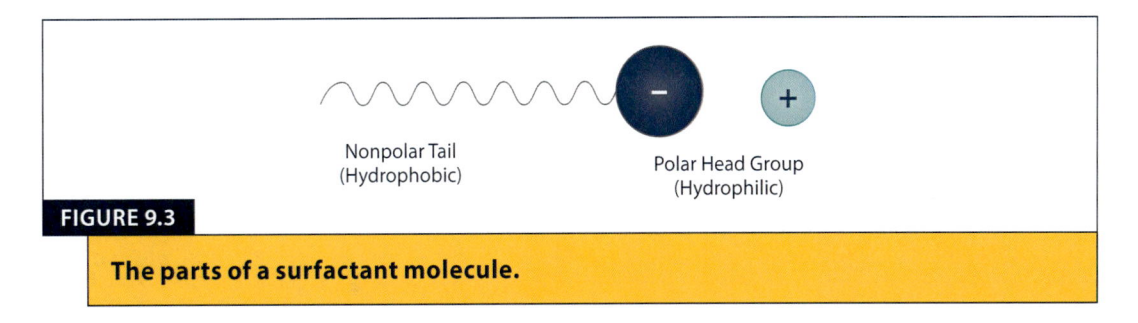

FIGURE 9.3

The parts of a surfactant molecule.

Soaps are sodium or potassium salts of fatty acids (carboxylic acids with long hydrocarbon chains), while detergents are usually sulfonate salts, which contain $-SO_3^-$ anions. The Na^+ or K^+ cation is only a spectator ion and has no role in the cleaning action.

FIGURE 9.4

Molecule chains of a soap and a detergent.

The hydrocarbon portion of these molecules, represented by the zig-zag line, is the nonpolar end and mixes with oily, greasy dirt. This hydrophobic chain mixes with the water only because the hydrophilic end pulls it into solution.

Certain substances will hinder the emulsifying power of these salts. Hard water contains large quantities of minerals such as Fe^{3+}, Mg^{2+}, and Ca^{2+} ions. Insoluble salts of the soaps (e.g., calcium stearate) precipitate in the presence of these ions, and these residues are difficult to rinse away. The iron, magnesium, and calcium ions do not precipitate with the sulfonates in detergents; they are water-soluble. This is one of the advantages of detergents over soaps.

When clothes are *dry cleaned*, a nonpolar liquid is used instead of water. The nonpolar dry cleaning fluids are good for dissolving away oil and grease, but would be ineffective in washing away ionic compounds such as table salt.

In the following study, make close observations of each solution or suspension. Keep in mind that the cleaning power of a soap or detergent depends on how well the surfactant is able to emulsify the dirt or grease. To emulsify means to break up an oily deposit into minute droplets which will be dispersed in the aqueous phase.

LABORATORY 9 SAPONIFICATION OF VEGETABLE OIL AND SOAP PROPERTIES

PROCEDURE

PART I. SYNTHESIS OF SOAP

1. Add 10 mL of the provided oil, 10 mL of ethyl alcohol, and 15 mL of 20% sodium hydroxide to a 250-mL beaker. Place the beaker on a hot plate at a medium high setting. Stir occasionally.

2. Once the solution heats significantly, the alcohol will start to boil. *At this point, continuously stir the mixture vigorously until the major reaction has taken place.* **Serious bumping** and spattering may occur if this precaution is not taken.

   As the saponification reaction proceeds, soap will precipitate out of the mixture. Continue stirring and heating for several minutes to evaporate the alcohol.

   FIGURE 9.5 Precipitate formed from heating the solution.

3. Remove the beaker from the heat. Add 40 mL of 20% sodium chloride solution and mix thoroughly. The salt solution will help to precipitate the soap.

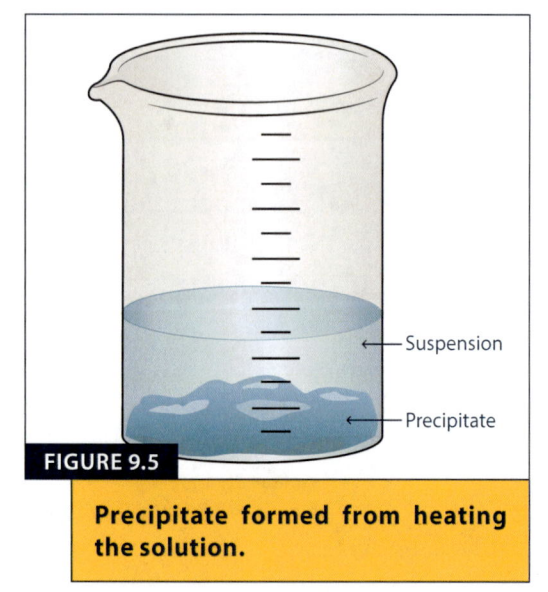

4. Collect the soap by vacuum filtration with a Buchner funnel. After the liquid has been pulled off, stop the suction. Add 10 mL of water and stir the mixture to wash the excess sodium chloride and sodium hydroxide from the solid. Remove the water by reapplying suction.

5. Transfer the soap to a paper towel. Investigate its characteristics. Make a few milliliters of soap solution and test with pH paper or litmus.

6. Wash the filter funnel and flask and return them to their original location.

LABORATORY 9 SAPONIFICATION OF VEGETABLE OIL AND SOAP PROPERTIES

PART II. SOAP, DETERGENTS, AND EMULSIFICATION

A. SOAP AND DETERGENT SOLUTIONS

1. Make a soap solution containing a half spoonful of *Ivory Snow*® or shavings of an *Ivory*® soap bar in 100 mL of deionized water.

2. Make a detergent solution containing a half spoonful of *All*® and 100 mL of deionized water.

3. Make a soap solution using your synthesized soap by mixing some with 100 mL of deionized water.

4. Mix each vigorously and then allow the undissolved particles to settle.

5. Pour the clear solution into a clean beaker or flask. Since you will be looking for the formation of a precipitate in some cases, these solutions must be relatively clear. If necessary, filter through a small bit of glass wool.

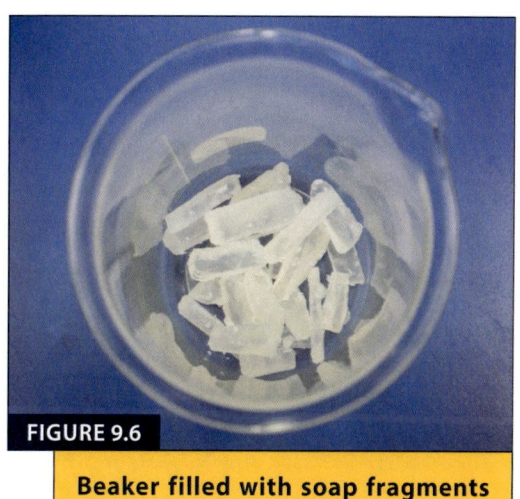

FIGURE 9.6 Beaker filled with soap fragments for creating the solutions.

B. EMULSIFICATION

1. In 4 large test tubes, place 5 mL of soap solution, 5 mL of detergent solution, 5 mL of your synthesized soap solution, and 5 mL of water, respectively.

2. Estimate the pH of each with pH indicator paper.

3. Add 10 drops of cooking oil to each and shake the test tube vigorously. Compare the emulsifying activity in each test tube. Which test tube has the largest oil droplets remaining? (Ignore the soap bubbles on top.)

4. Pour the solution out of each test tube. Gently fill each test tube with tap water and empty twice. Which tubes are clean and free of oil droplets?

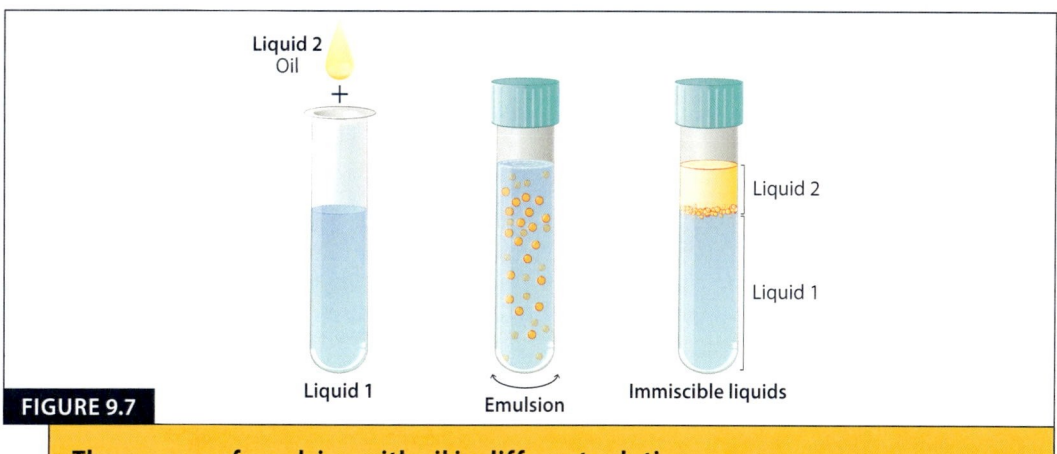

FIGURE 9.7 The process of emulsion with oil in different solutions.

INTRODUCTORY CHEMISTRY LABORATORY MANUAL 71

LABORATORY 9 SAPONIFICATION OF VEGETABLE OIL AND SOAP PROPERTIES

C. EFFECTS OF CATIONS OFTEN FOUND IN GROUND WATER

1. In 3 large test tubes, place 5 mL of soap solution, 5 mL of your synthesized soap solution, and 5 mL of detergent solution, respectively.

2. Add 2 mL of 1% Ca^{2+} solution (aqueous $CaCl_2$) to each and shake. What do you observe? Is a precipitate formed?

3. Pour out the contents and rinse the test tubes as you did in Part A. Are the test tubes clean, or do they contain a residue?

4. Repeat by adding 2 mL of 1% Na^+ solution (aq. NaCl) to fresh soap and detergent solution samples.

5. Repeat with 2 mL of 1% Fe^{3+} (aq. $FeCl_3$).

Which cations (Fe^{3+}, Na^+, or Ca^{2+}) would interfere with the washing process?

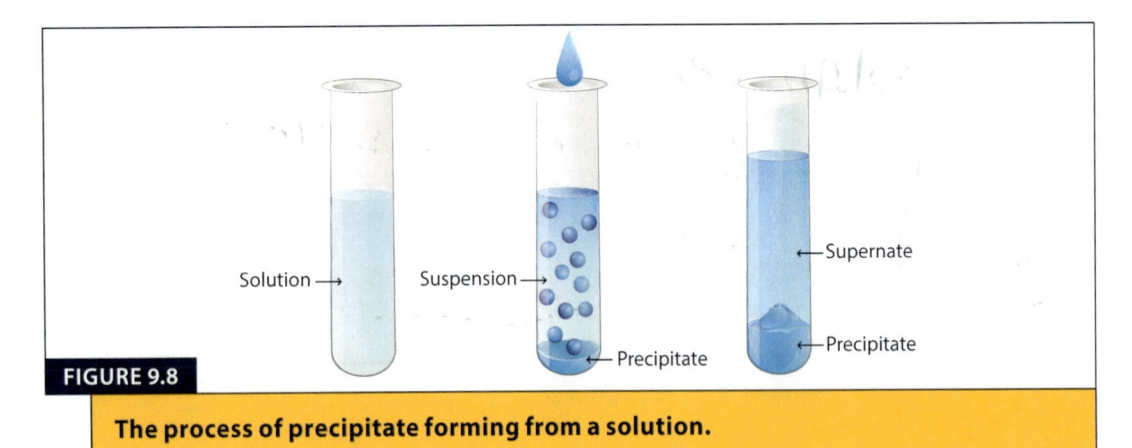

FIGURE 9.8 The process of precipitate forming from a solution.

LABORATORY 10

SYNTHESIS OF ASPIRIN

OBJECTIVES

- Use a chemical equation to relate mass quantities of reactant and product.
- Crystallize a product from solution.
- Use vacuum filtration to collect a solid product.
- Apply theoretical yield and percent yield to a laboratory reaction.
- Characterize products of a chemical reaction.
- Identify functional groups and structural features of an organic compound.
- Evaluate aspects of commercial preparation of chemical products.

PART I. SYNTHESIS OF ASPIRIN

INTRODUCTION

Aspirin is the most widely used drug in the world with annual consumption of 100 billion tablets. Aspirin is a drug with analgesic (pain reliever), antipyretic (fever reducer), and anti-inflammatory (reducing inflammation, swelling or arthritic symptoms) properties.

The use of aspirin-like compounds had its origin in folk medicine. Chewing on willow bark was known to relieve pain and reduce fever. Salicilin was extracted from willow bark in 1763, but its use was unpleasant due to its taste. Modifications of salicylic acid were tried. Aspirin was first synthesized in 1853 by Charles F. Gerhardt, a German chemist, as a byproduct of coal tar. Felix Hoffman investigated its medicinal value in 1899. In that year, the Bayer Company patented the synthesis of aspirin and sold the first commercial aspirin.

Aspirin, the common name for acetylsalicylic acid, $C_9H_8O_4$, is a white, bitter solid. It can be prepared by reacting acetic anhydride with salicylic acid. The figure shows the structural changes that occur during the reaction. The –OH (alcohol) group of the salicylic acid combines with the O=CCH$_3$ (acetyl) group of acetic anhydride to form

INTRODUCTORY CHEMISTRY LABORATORY MANUAL **77**

acetylsalicylic acid. Acetic acid is a byproduct. Aspirin is fairly insoluble in cold water; however, a small amount (0.25 grams) will dissolve in 100 mL of cold water. This insolubility allows aspirin to be crystallized from water and collected by filtration.

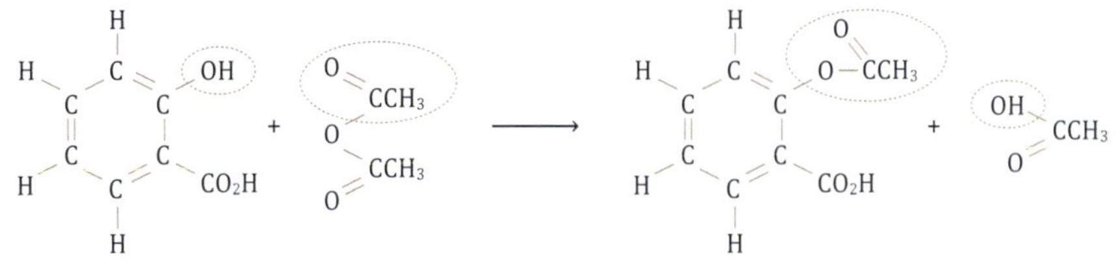

Salicylic Acid $C_7H_6O_3$ **Acetic Anhydride** $C_4H_6O_3$ **Acetylsalicylic Acid** $C_9H_8O_4$ **Acetic Acid** $C_2H_4O_2$

FIGURE 10.1

Structural changes that occur during the reaction of acetic anhydride with salicylic acid.

The synthesis of aspirin can be represented by the following equation:

$$C_7H_6O_3 + C_4H_6O_3 \rightarrow C_9H_8O_4 + C_2H_4O_2.$$

While less informative about the structural changes that occur in the reaction, this equation indicates the mass relations that are represented by the equation. Note that the number of atoms of each element in the reactants is equal to the number of atoms of each element in the products. A fundamental law of chemistry is that mass is conserved during a chemical reaction.

The total mass of all of the products is equal to the total mass of all reactants. In addition, the formulas represent the relative masses of the individual reactants and products involved in the reaction. Using the number of atoms in the formula and the atomic masses, the formula masses of salicylic acid and of aspirin are determined as indicated in the table below.

TABLE 10.1

SALICYLIC ACID			ASPIRIN		
ELEMENT	# ATOMS	MASS	ELEMENT	# ATOMS	MASS
C	7	7 × 12.011 = 84.077	C	9	9 × 12.001 = 108.009
H	6	6 × 1.008 = 6.048	H	8	8 × 1.008 = 8.064
O	3	3 × 15.999 = 47.997	O	4	4 × 15.999 = 63.996
	Total Mass =	138.122		Total Mass =	180.069

The analysis in the table indicates that 180.069 g of aspirin will be produced when 138.122 g of salicylic acid reacts. These values can be used to generate a conversion factor by dividing:

$$\frac{180.069 \text{ g aspirin}}{138.122 \text{ g salicylic acid}} = \frac{1.30 \text{ g aspirin}}{1.00 \text{ g salicylic acid}}.$$

Thus, the amount of aspirin that can be produced from a given mass of salicylic acid can be calculated.

In this experiment, aspirin will be synthesized and the percent yield of aspirin will be determined. Percent yield defines the relative amount of product actually obtained compared to the maximum which could be obtained if the limiting reactants were converted entirely to product and none of the material was lost in handling. Percent yield is defined as follows:

$$\% \text{ yield} = \frac{\text{actual yield}}{\text{theoretical yield}} \times 100\% = \frac{\text{mass of aspirin obtained}}{\text{mass of aspirin possible}} \times 100\%.$$

PROCEDURE

1. Prepare a water bath by using a hot plate to heat approximately 200 mL of tap water in a 400-mL beaker. Also, place a wash bottle of deionized water in an ice bath in an 800-mL beaker. (Ice baths are prepared by filling a beaker half full of ice and adding tap water to the height of the ice.)

2. Weigh out approximately 3 grams of salicylic acid on a piece of weighing paper by taring the balance with the weighing paper on it and then adding salicylic acid. Be sure to record the exact mass. Place the salicylic acid in a 125-mL Erlenmeyer flask.

3. At the hood, first add 6.0 mL of acetic anhydride, using the dropper pipette provided to the flask, then add 8 drops of concentrated (18 M) sulfuric acid using the dropper pipette provided.

CAUTION	Both are very corrosive to the skin. Wear gloves and dispose of them after handling these chemicals. Clean up any spills.

Mix gently by swirling.

4. Heat the reaction mixture in a warm water bath (80–90 °C) for 10 minutes while swirling occasionally. All of the white salicylic acid should dissolve. Remove the flask from the heat and let it cool until it approaches room temperature. After some initial cooling, place it in an ice water bath to speed up the cooling.

5. While waiting for the sample to cool, weigh a dry watch glass.

6. If there are no crystals after cooling, scrape the bottom of the flask to induce crystallization.

LABORATORY 10 SYNTHESIS OF ASPIRIN

7. Once crystals form, add 40 mL of the cold deionized water to the flask, and then cool it in an ice water bath. Stir the material for several minutes and break up any clumps of crystals.

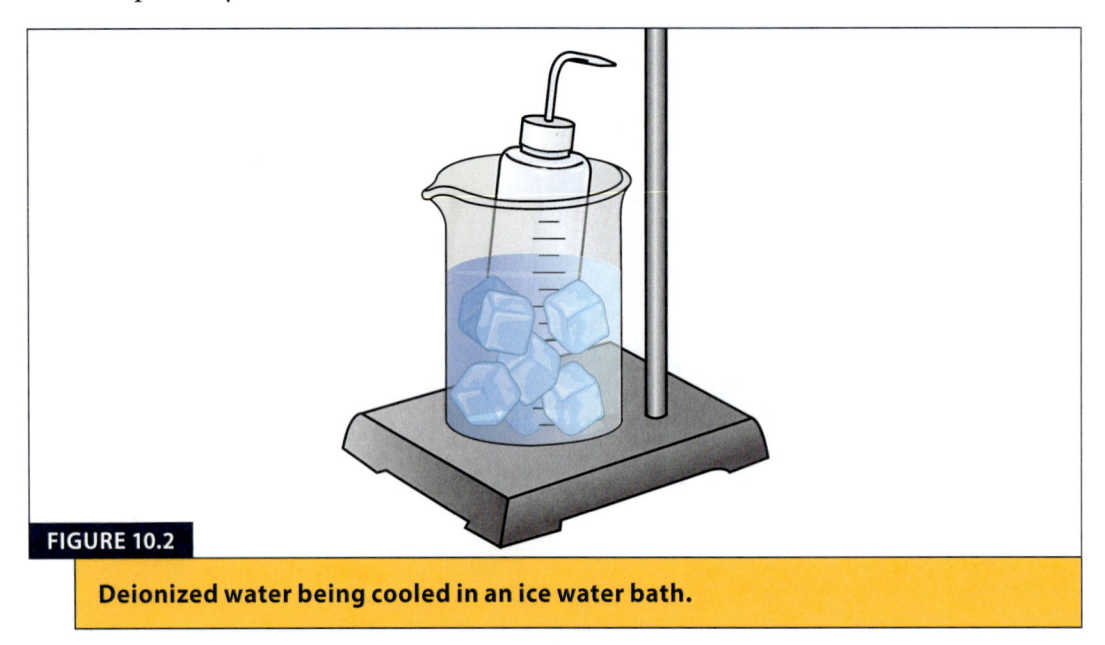

FIGURE 10.2
Deionized water being cooled in an ice water bath.

8. Set up the Buchner funnel, secured with an iron ring. Place a piece of filter paper in the funnel and wet it with water.

9. Filter out the crystals by suction filtration using a clamped Buchner funnel with a dampened piece of filter paper. Wash the crystals with a little ice cold water. Refilter the filtrate if a lot of crystals are suctioned into filter flask. Leave the suction on while completing *Part II. Formation of Methyl Salicylate.*

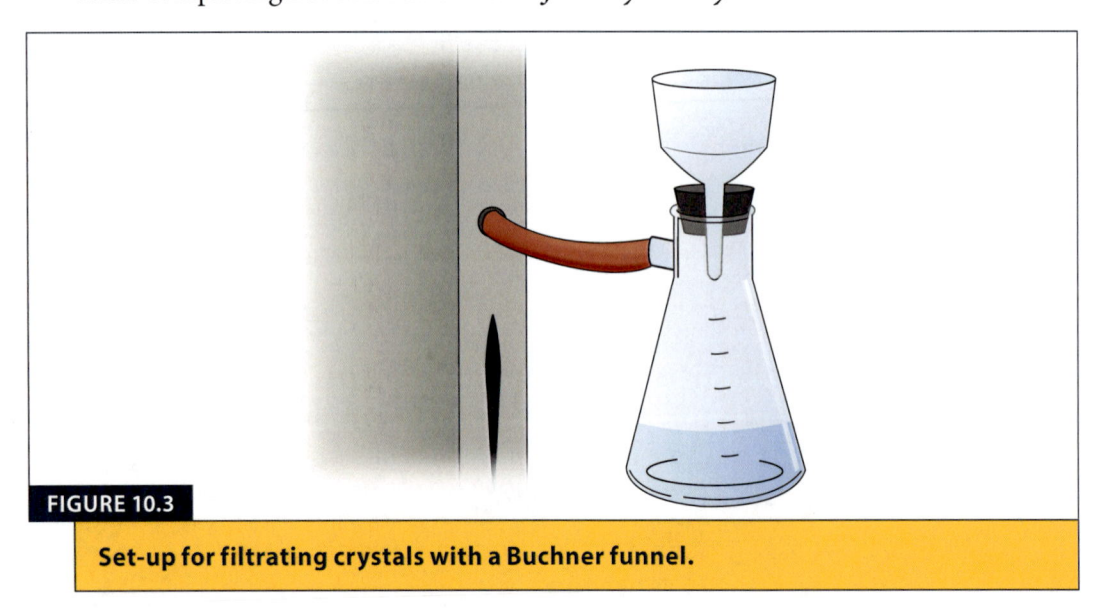

FIGURE 10.3
Set-up for filtrating crystals with a Buchner funnel.

10. Determine the mass of your product by using a preweighed watchglass. (You may store your aspirin until your next meeting in order for it to dry thoroughly. Follow your instructor's guidelines.)

LABORATORY 10 SYNTHESIS OF ASPIRIN

PART II. FORMATION OF METHYL SALICYLATE

INTRODUCTION

An ester forms from the combination of a carboxylic acid (–CO$_2$H) group and an alcohol (–OH) group. Since salicylic acid contains both an acid functional group and an aromatic alcohol (a phenol) functional group, it is possible for salicylic acid to be either the acid part or the alcohol part of an esterification reaction.

In the synthesis of aspirin, the alcohol functional group of salicylic acid reacted with acetic anhydride (an "activated" derivative of a carboxylic acid) to produce an ester.

The carboxylic acid functional group of salicylic acid (circled) could also be used to form an ester by reacting with an alcohol. In this experiment, the carboxylic acid will react with methanol to form an ester- methyl salicylate- and water.

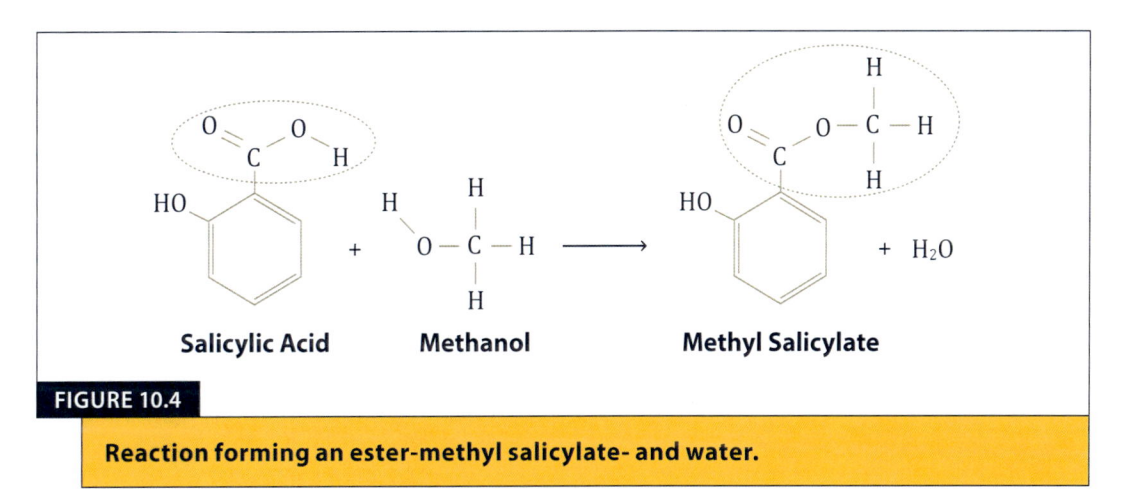

FIGURE 10.4

Reaction forming an ester-methyl salicylate- and water.

PROCEDURE (WORK WITH A PARTNER)

1. Add 6 drops of methanol to a medium-sized test tube.
2. Add an amount of salicylic acid equivalent to the size of the head of a match (about 0.1 g).
3. At the hood, add 1 drop of concentrated sulfuric acid. *(CAUTION!)*
4. Use a test tube holder to hold the sample while it is heated in a boiling water bath for a few minutes.
5. Add about 15 drops of water. Cautiously smell the methyl salicylate product. What does it smell like?

CLEAN-UP AND WASTE DISPOSAL

Dispose of the methyl salicylate product by flushing it down the drain.

LABORATORY 11

IDEAL GAS LAW MOLECULAR WEIGHT OF A VAPOR

OBJECTIVES

- Apply the Ideal Gas Law to a scientific problem.
- Determine the molar mass of a volatile liquid.
- Apply concepts relating the composition of a chemical substance to its formula.

INTRODUCTION

There are several methods by which the molecular weight of a compound can be determined in the laboratory. In this experiment you will use a method that can be applied to volatile liquids. Volatile liquids are those liquids that have a moderate to high vapor pressure at or near room temperature. Most of these types of compounds will behave like an ideal gas when converted to the vapor state. This means that the **ideal gas law** will apply:

$$PV = nRT.$$

In this equation, **P** is the **pressure** of the gas, **V** is the **volume** of the gas, **n** is the **amount** of the gas in moles, and **T** is the **Kelvin temperature** of the gas. **R** is called the **ideal gas constant.** The value of R will differ depending on the units used for pressure and volume. When P is in atmospheres and V is in liters, the value of R is **0.08206 (L atm)/(mol K).**

This equation is useful because it allows one to calculate the pressure, volume, temperature, or number of moles of a gas simply by knowing the other three variables and doing a little algebra.

In the following experiment you will use a setup with which you can easily determine the values for pressure, volume, and temperature of a gas. Once these values have been found, you can determine the amount of the gas in moles, and from the mass of the gas in grams, you can calculate the molar mass (molecular weight) of the gas as follows:

$$\text{Moles of gas} = n = \frac{PV}{RT} \text{ and molecular weight} = \text{grams of } \frac{\text{grams of gas}}{\text{moles of gas}}.$$

INTRODUCTORY CHEMISTRY LABORATORY MANUAL **85**

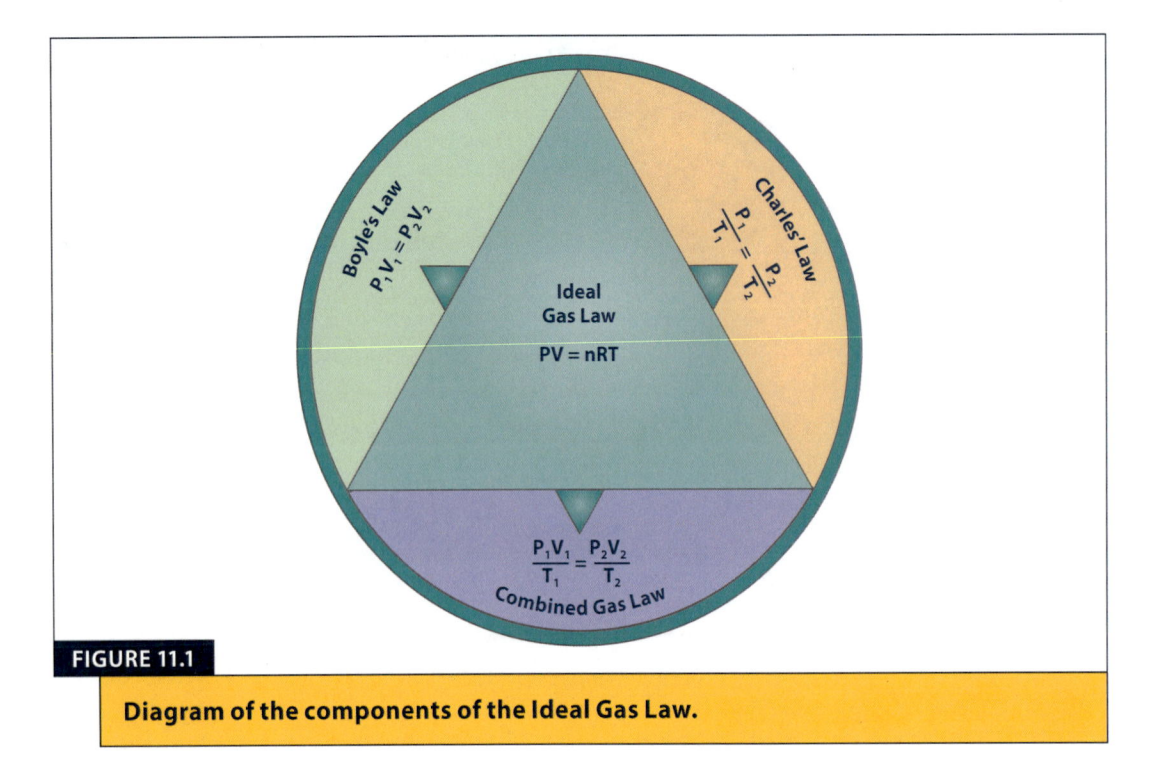

LABORATORY 11 IDEAL GAS LAW: MOLECULAR WEIGHT OF A VAPOR

FIGURE 11.1

Diagram of the components of the Ideal Gas Law.

Gases, unlike solids and liquids, have neither fixed volume nor shape. They expand to fill the entire container in which they are held. The pressure of a gas is defined as force per area. The standard or SI unit for pressure is the **Pascal** (Pa) which is the force exerted by the gas in Newtons divided by area in square meters, N/m^2. However, **atmospheres** (atm) and several other units are also commonly used. The table below shows the conversions between these units:

TABLE 11.1

COMMON UNITS OF PRESSURE
1 Pascal (Pa) = 1 Newton per square meter = 1 N/m^2
1 atmosphere (atm) = 1.01325×10^5 Pa = 101.325 kPa
1 bar = 105 Pa
1 atm = 760 torr = 760 mmHg (millimeters of mercury)
1 atm = 29.92 inHg (inches of mercury)
1 atm = 14.70 psi (pounds per square inch)

The ideal gas law assumes several factors about the molecules of gas. The volumes of the gas molecules themselves are considered negligible compared to the volume of the container in which they are held. We also assume that gas molecules move randomly and collide in completely elastic collisions. Attractive and repulsive forces between the molecules are, therefore, considered negligible.

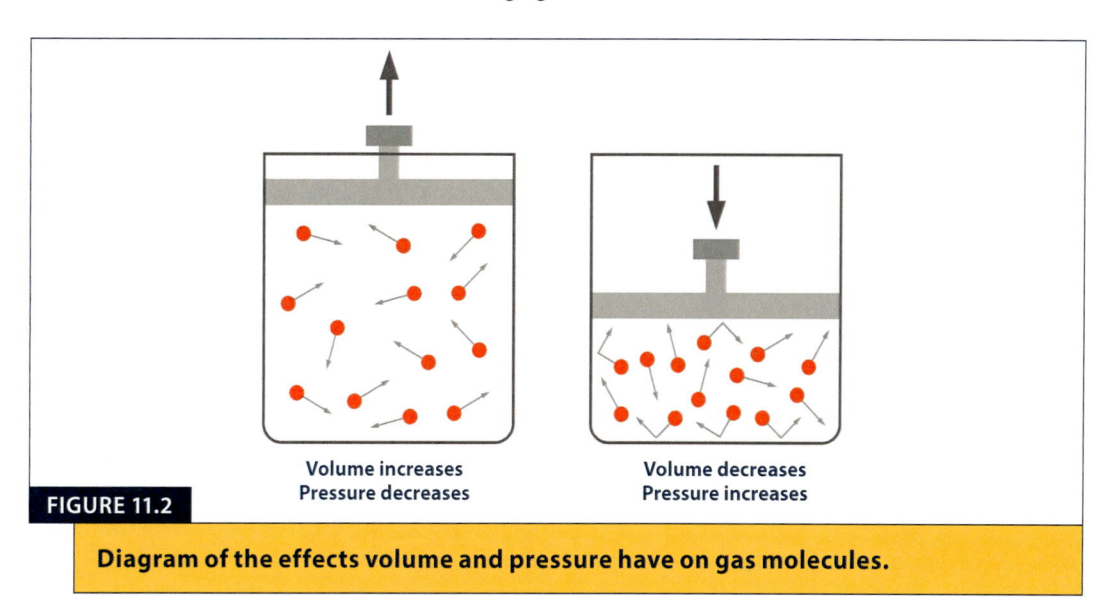

FIGURE 11.2

Volume increases
Pressure decreases

Volume decreases
Pressure increases

Diagram of the effects volume and pressure have on gas molecules.

We can also use the ideal gas law to quantitatively determine how changing the pressure, temperature, volume, and number of moles of substance affects the system. Because the gas constant, R, is the same for all ideal gases in any situation, if you solve for R in the ideal gas law and then set two terms equal to one another, you obtain a convenient equation called the **combined gas law**:

$$R = \frac{P_1 V_1}{n_1 T_1} = \frac{P_2 V_2}{n_2 T_2}$$

where the values with a subscript of "1" refer to initial conditions and values with a subscript of "2" refer to final conditions.

SAMPLE CALCULATION

To calculate the molecular weight of a volatile liquid, the liquid was vaporized in an Erlenmeyer flask which had a *total* volume of 152 mL. In the procedure, the flask containing an excess amount of the volatile liquid was covered with aluminum foil with a tiny pinhole, and then the flask and the liquid were placed in a boiling water bath at 100 °C. The atmospheric pressure was measured 754 torr with a barometer in the room.

As the liquid in the flask vaporized, the excess vapor escaped through the pinhole until no visible liquid remained in the flask. The flask was then removed from the water bath and allowed to cool. The mass of the flask, foil, and vapor was 94.53 g. The initial mass of the dry, empty flask and foil was 94.12 g.

LABORATORY 11 IDEAL GAS LAW: MOLECULAR WEIGHT OF A VAPOR

From this information, calculate the molecular weight of the volatile liquid in grams per mole.

ANSWER: The calculation is actually quite straightforward, but we must take care to use the correct *units* on all numbers. Pressure should be in atmospheres, volume in liters, and, of course, temperature in Kelvin. The calculation consists of two parts:

1. Use PV = nRT to calculate the moles of gas, n.
2. Divide the grams of gas by the moles of gas (calculated in Step 1) to obtain the molecular weight of the gas in grams per mole.

- $n = \frac{PV}{RT}$.

$$P = \frac{754 \text{ torr}}{1} \times \frac{1 \text{ atm}}{760 \text{ torr}} = 0.9921 \text{ atm.}$$

$$V = \frac{152 \text{ mL}}{1} \times \frac{1 \text{ L}}{1000 \text{ mL}} = 0.152 \text{ L.}$$

$$T = 100\ °C + 273.15 = 373.15 \text{ K.}$$

$$n = \frac{(0.9921 \text{ atm})(0.152 \text{ L})}{0.08206 \text{ L atm/mol K}(373.15 \text{ K})} = 0.004325 \text{ mol.}$$

- The mass of a gas = 94.53 g − 94.12 g = 0.41 g.

$$\text{The molecular weight of the gas} = \frac{0.41 \text{ g}}{0.004325 \text{ mol}} = 83 \text{ g/mol.}$$

PROCEDURE

1. Determine the combined total mass of a dry 125-mL Erlenmeyer flask and a square piece of aluminum foil, and record this mass on the report form.

2. Create a hot water bath by filling a 600-mL beaker half full of water. Heat to boiling on a hot plate, and while waiting for the water to boil, use a 10-mL graduated cylinder to measure and pour approximately 2 mL of your unknown liquid sample into the flask. (The liquid is colored by a trace of I_2 to make it more visible.) Secure the aluminum foil over the mouth of the flask. Make a *tiny* hole in the foil with a pin to let excess vapor escape during heating.

3. Clamp the flask assembly into the beaker so that flask is as far down as possible in the beaker. Heat at the

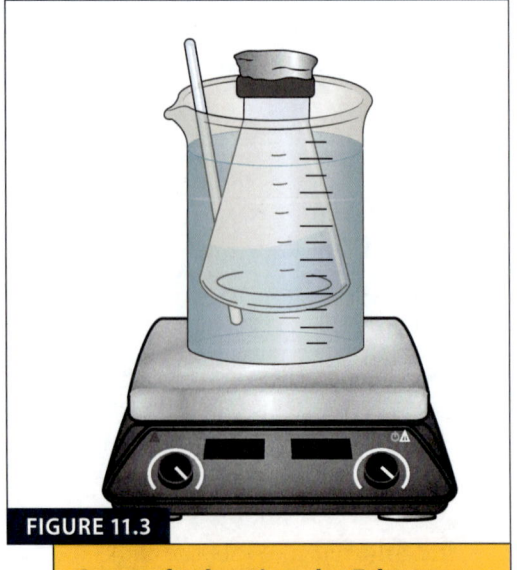

FIGURE 11.3
Set-up for heating the Erlenmeyer flask in a beaker.

LABORATORY 11 IDEAL GAS LAW: MOLECULAR WEIGHT OF A VAPOR

boiling point of water until liquid is no longer visible in the flask. Try not to heat it too much past this point as your gas vapor will start to be replaced by air. Measure the temperature of the gas occupying the flask by recording the temperature of the boiling water surrounding the flask. Since the flask is even so slightly open to the atmosphere, the pressure of the gas must be equal to the barometric pressure. Measure the atmospheric pressure with the barometer in the laboratory and record the current barometric pressure on the Report Form.

4. Remove the flask from the water bath by loosening the clamp from the supporting rod and moving it upward. Allow the flask to cool to room temperature, and dry it on the outside gently and thoroughly with a paper towel. Weigh the flask along with its contents and the aluminum foil, and record this value on the Report Form.

5. Accurately measure the volume of the flask (it is more than 125 mL!) by filling it with water all the way to the brim and measuring the volume of water with a 100-mL graduated cylinder. Record the flask volume on the Report Form.

6. Calculate the moles of vapor from its pressure, volume, and temperature, and from this value and the mass of the vapor, calculate the molar mass, or molecular weight, of the vapor.

$$R = \frac{P_1 V_1}{n_1 T_1} = \frac{P_2 V_2}{n_2 T_2}$$

P - Pressure   V - Volume (L)   n = moles

R = Constant = 0.08206 (L atm/mol K)

T = Temp = K

°C + 273.15 = K

LABORATORY 12

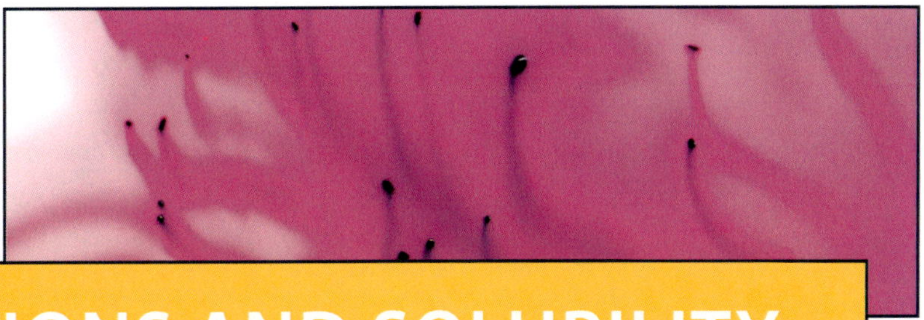

SOLUTIONS AND SOLUBILITY

OBJECTIVES

- Define the nature of solutions.
- Experimentally observe and test unsaturated, saturated, and supersaturated solutions.
- Observe the characteristics of polar and nonpolar liquids.
- Evaluate the ability of polar and nonpolar solvents to dissolve common materials.
- Determine the differences in interaction between a dye and polar and nonpolar fabric fibers.

INTRODUCTION

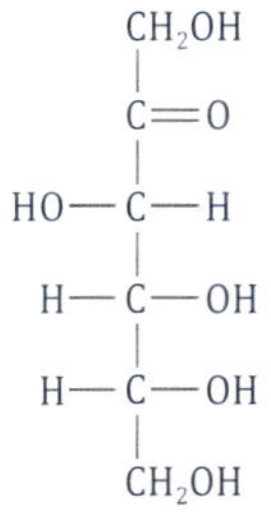

Solutions are an intimate part of our daily experiences. Phrases like "oil and water don't mix," the use of furniture polish to protect wood, and dry cleaning all reflect the nature of chemical substances and their ability to mix, or form solutions. The term *solution* refers to a homogenous mixture of two or more chemical substances which are intermixed on the atomic, molecular or ionic level. The most familiar solutions are liquids, but gaseous and solid solutions are also possible. Air and metal alloys are examples, respectively. In this experiment, discussion will be limited to liquid solutions.

FIGURE 12.1 Stucture of a fructose molecule.

To consider the molecular picture of solutions, the nature of chemical substances needs to be considered. Compounds are usually classified as either molecular covalent or ionic. The sugar fructose, $C_6H_{12}O_6$, is a substance that is classified as **molecular.** Its atoms are bonded into an uncharged discrete unit. Water, H_2O, is also molecular.

LABORATORY 12 SOLUTIONS AND SOLUBILITY

FIGURE 12.2
Model of an ionic compound.

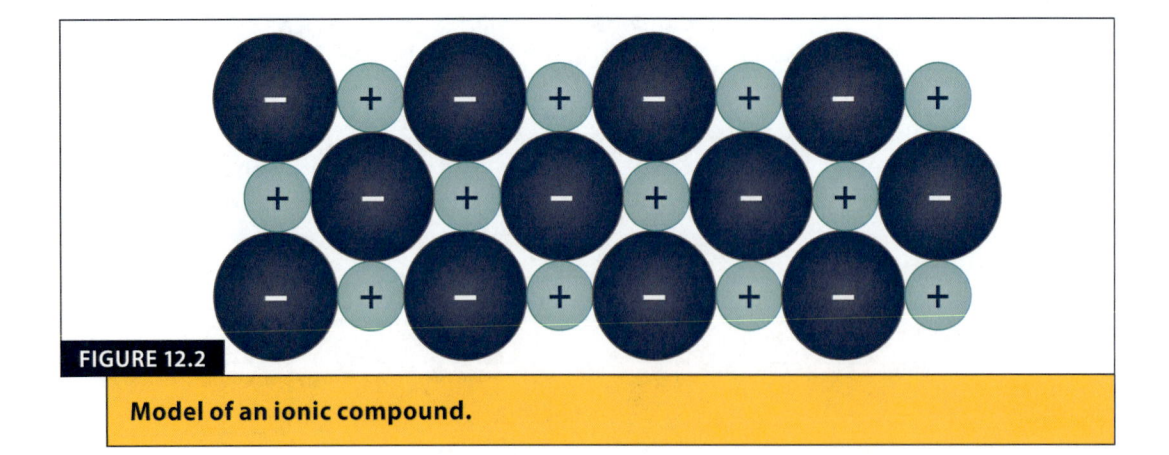

Another class of compounds is called ionic. **Ionic compounds** consist of positively charged cations and negatively charged anions. The cations and anions are separate, unique species, but may be held together by electrostatic attraction. In the solid state, ionic compounds exist as arrays of cations and anions. No discrete cation-anion unit exists.

When atoms are bonded in covalent molecules, the positions of the negatively charged electrons around the atoms may be displaced enough to cause one end of the molecule to have a slight negative charge. The other end of the molecule would then have a slight positive charge. This makes the overall molecule **polar.** Polar molecules like water have uneven distribution of negatively charged electrons. Thus, the negative charge may dominate on one side of the molecule (δ–) while the other side has some residual positive charge (δ+). This is fairly obvious for a molecule like HCl where the two ends of the molecule are different. By having partially charged ends, polar molecules exhibit electrostatic interaction with other polar or charged substances.

If electrons are uniformly distributed across the molecule, it will be **nonpolar.** Thus, I_2 (I—I) is nonpolar, and HI (H—I) is polar.

NATURE OF SOLUTIONS

Consider what happens when sugar (sucrose) is mixed with water; the sugar **dissolves** (not "melts") to form a solution. Individual sugar molecules are removed from solid sugar crystals as the solid dissolves. In the solution, each sugar molecule is separated from all other sugar molecules by many water molecules. The dissolving medium, water, is considered the **solvent.** The material that dissolves, sugar, is the **solute.** In some solutions, such as in your car's cooling system, the two substances (in this case, water and ethylene glycol) are both liquids and are present in nearly the same amounts. The definitions of solute and solvent become blurred in this case.

LABORATORY 12 SOLUTIONS AND SOLUBILITY

FIGURE 12.3

Model of Solution of an Ionic Compound in Solvent

Model of Solution of a Molecular Compound (sucrose) in Solvent

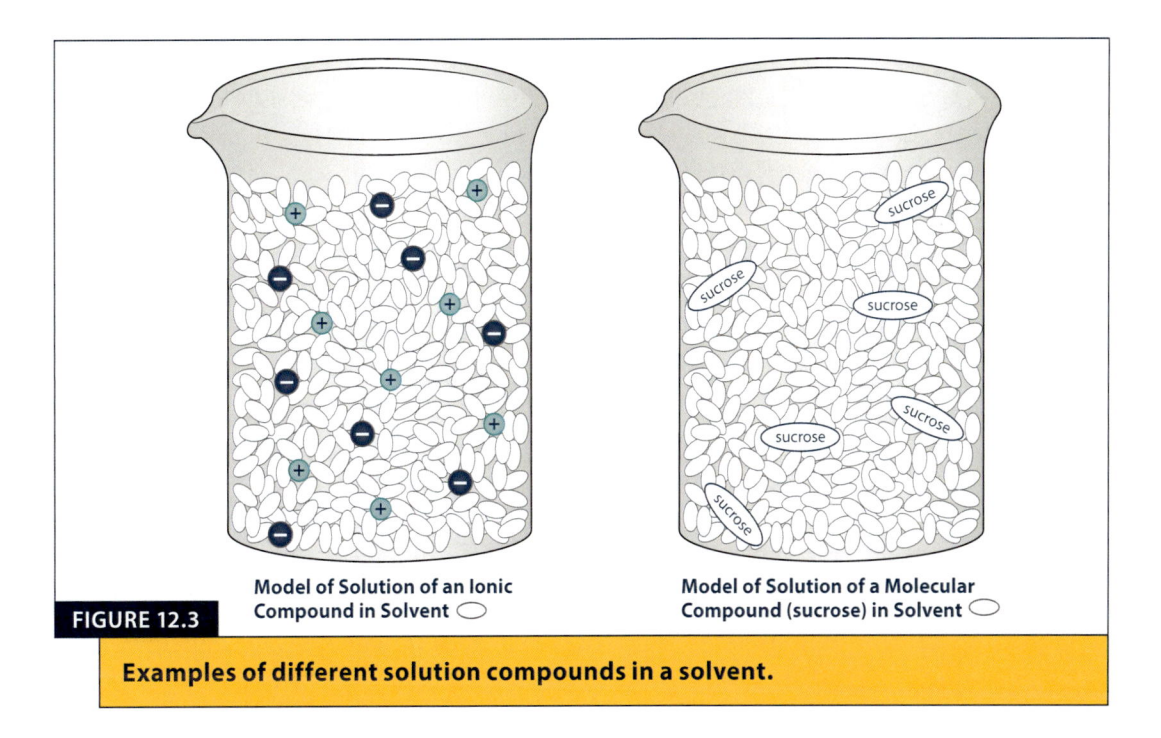

Examples of different solution compounds in a solvent.

Table salt dissolves readily in water, but does so differently than sugar. Table salt is crystalline ionic sodium chloride, NaCl, and exists as an ordered 3-dimensional array of alternating sodium cations, Na^+, and chloride anions, Cl^-, as indicated on the previous page. When ionic compounds dissolve in water, the individual ions become separated by water molecules. For salt, each sodium ion is surrounded by many water molecules and is well separated from other sodium ions and from the chloride ions when in solution. Because these positive and negative ions are free to move through the solution, thus carrying charge, solutions of salts are good conductors of electricity. Solutions of dissolved salts are called electrolytes because they are capable of conducting electricity. The sugar molecules, which are electrically neutral, carry no charge. Solutions of molecular substances are nonelectrolytes.

The division of substances into ionic and molecular covalent will be handy when discussing solubility.

SOLUBILITY

All salts are soluble in water to some extent. The range of this solubility phenomenon is vast. Two pounds of silver nitrate will dissolve in only a half cup of water. Only a few "units" (picograms, 10^{-12} g) of silver sulfide will dissolve in a bathtub full of water to form a saturated solution of Ag_2S. Why some salts are readily soluble in water and others only sparingly is well-understood, but we will not go into the thermodynamics of this issue.

INTRODUCTORY CHEMISTRY LABORATORY MANUAL

Polar molecules tend to mix with other polar molecules, but nonpolar substances do not mix well with very polar substances like water. Nonpolar molecular substances such as hydrocarbons, fats, and esters, to name a few, do not dissolve in polar water because they do not interact with water effectively.

The terms **saturated** and **unsaturated** refer to the relative amounts of solute in solution *compared to the maximum amount of solute that the solvent can dissolve at a certain temperature.* A good way to test the condition of a solution is to add a few small crystals of the solute. By definition, an *unsaturated* solution can dissolve more solute, and the small crystals will dissolve. A *saturated* solution will not dissolve the crystals because the solvent already supports all the solute it can. Obviously, a situation where there is undissolved solute present would indicate that the solution is already saturated (assuming the solute was given enough time to dissolve). A **supersaturated solution** is an unstable condition where more solute is dissolved than should be possible. When the small crystal is added, it will not dissolve and, in fact, it will cause all of the excess solute to crystallize out of solution until a saturated solution exists.

INTERMOLECULAR FORCES AND DISSOLVING

Only a few organic compounds readily dissolve in water. Those that do are members of certain classes of compounds—sugars, alcohols, glycols, and low molecular weight carboxylic acids, like acetic acid. All of these compounds have one or more –OH groups as part of the molecule. A special type of intermolecular force is acting here between the solvent molecules, H—O—H, and solute molecules containing –OH or –NH$_2$ groups. This force is termed **hydrogen bonding** and involves the formation of a strong attraction between a hydrogen atom of an –OH or –NH group on one molecule and an O or N atom in another nearby molecule. Hydrogen bonding is the interaction that makes water a unique and unusual liquid. Hydrogen bonding is also what holds two DNA strands together. Hydrogen bonding is not a formal chemical bond but it is an unusually strong interaction between the two species involved and plays an important part in the dissolution of many compounds.

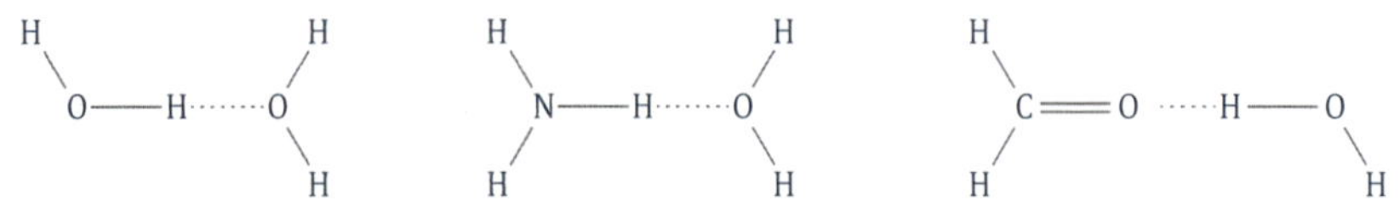

FIGURE 12.4

Hydrogen bonding of molecular structures.

LABORATORY 12 SOLUTIONS AND SOLUBILITY

The above classifications (ionic, hydrogen bonding, polar covalent, or nonpolar covalent) are used to predict the solubility of various substances with the rule "like dissolves like." In general, the closer the nature of a substance is to that of the solvent, in the order listed below, the more likely it is to be soluble.

$$\text{Ionic : Hydrogen Bonding : Polar : Nonpolar}$$

PROCEDURE

PART I. UNSATURATED, SATURATED, AND SUPERSATURATED SOLUTIONS

A. SODIUM ACETATE TEST

1. Add 2 red plastic spoons (about 5 g) of sodium acetate ($NaC_2H_3O_2$) to the bottom of a large clean, dry, labeled test tube. Line the test tube with a rolled-up piece of paper to avoid leaving any crystals on the test tube wall; remove the paper when the transfer is complete.

2. Add 20 mL of deionized water and heat the test tube in a water bath with occasional swirling until the solid is completely dissolved.

3. Set the test tube aside to cool undisturbed. If any solid crystallizes, repeat the heating. (Continue with other experiments as you wait.)

4. Record the appearance of the solution after it has cooled about 30 minutes.

5. Feel the bottom of the test tube.

6. Add one crystal of sodium acetate. Observe what happens and note the temperature of the test tube.

7. Stir the contents to determine if they are completely solid.

B. POTASSIUM CHLORIDE

1. Put 5 mL of deionized water in a large test tube.

2. Put one rounded spoon (about 2 g) of KCl in a small beaker for later use.

Condition A

3. Add approximately ¼ of a spoonful of the KCl to the test tube and mix for a couple of minutes. Does the salt dissolve?

Add another small crystal of KCl. Does it dissolve? Was the solution unsaturated or saturated

INTRODUCTORY CHEMISTRY LABORATORY MANUAL 97

LABORATORY 12 SOLUTIONS AND SOLUBILITY

Condition B

4. Add the remaining KCl and stir. Does it all dissolve on mixing?

Condition C

5. Heat the test tube in a boiling water bath for 5 minutes, stirring occasionally. Now what's the situation? How could you test your prediction?

Condition D

6. Cool to room temperature. What do you observe?

   After cooling to room temperature, is the solution unsaturated or saturated? How could you test your answer?

PART II. SOLUBILITY AND STRUCTURE (WORK IN PAIRS)

The solubility of several substances will be examined in water, a polar solvent, and in hexane, a nonpolar solvent. The type of interactions are typical of those observed as you use liquids everyday. Do you wash a greasy pan in just water? Here's your chance to learn why that doesn't work.

It is suggested that one partner in each pair test the solubility of each material in water while the other partner tests the substance's solubility in hexane. Jointly, compare and record your careful observations and try to arrive at the nature of the substance. *Decide whether each substance is*

- *ionic,*
- *hydrogen bonding covalent,*
- *polar covalent* (capable of mixing with water), or is it
- *slightly polar covalent or*
- *nonpolar covalent* (capable of mixing with hexane)?

1. Add about 3 mL of water to 3 mL of hexane (C_6H_{14}) in a large test tube. (Make a mental note of what 3 mL looks like to save measuring this amount each time.) Mix well. What do you observe?

Waste Disposal

Discard all tests involving hexane into the waste bottle in the hood.

2. Take a small amount of solid iodine (about the size of a pin head) and try to dissolve it in 3 mL of water.

3. Repeat with 3 mL of hexane. Again, mixing is important to assure the substance is actually insoluble in the solvent rather than just slow to dissolve.

4. Repeat the tests just described in (2) and (3) for at least 5 substances provided.

FIGURE 12.5
Solid Iodine

PART III. DYES AND DYEING (WORK INDIVIDUALLY)

The dyeing of fabrics is another example of the action of attractive forces between molecules. When the molecular structures of the dye and the fabric are such that they are strongly attracted to each other, the dye will be colorfast and will not wash out of the fabric.

A typical dye is Congo Red, whose structure is shown below. (Note that in this style of representation, the intersection of two lines represents a carbon atom. The chemical formula for Congo Red is $C_{32}H_{22}N_6O_6S_2^{2-}$.) The important sites on the molecule are $-SO_3^-$ groups (anionic) and the hydrogen bonding of polar $-NH_2$ groups, whereas the rest of the molecule is nonpolar.

FIGURE 12.6
Structure of a Congo Red molecule.

LABORATORY 12 SOLUTIONS AND SOLUBILITY

Alizarin is a quinone dye used since antiquity. It is produced in the root of the madder plant. The red of the British "Red Coats" resulted from dyeing with alizarin. Madder, like indigo, was a substantial agricultural crop until organic chemists determined its molecular structure. Commercial synthesis soon replaced the much more expensive farming practice.

In this procedure you will investigate interactions between the dye molecule and several polymers by dyeing a strip of material that consists of the 13 different fabrics.

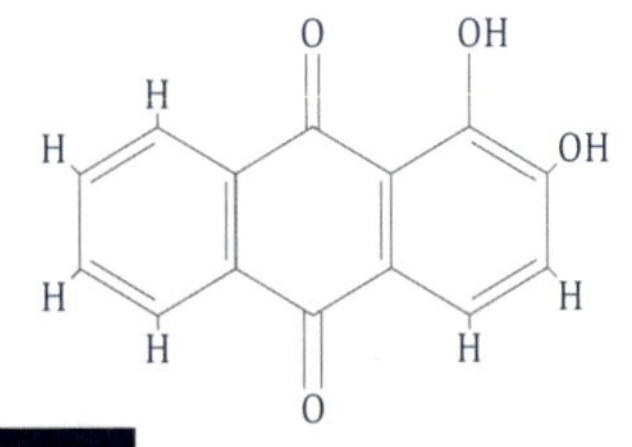

Threads are made from fibers, and fibers are made from strands of polymers. These long polymer strands may or may not have polar sites (various functional groups) along the polymer chain.

FIGURE 12.7 Molecular structure of Alizarin.

Cotton, linen and viscose (regenerated cotton, also called rayon) have polar –OH groups sticking out along the entire length of the polymer.

Hair, wool, and silk (all proteins) have several types of functional groups all along the polymer. Some of these groups react with acid sites, and some with base sites on a dye molecule.

Nylons and polyesters have fewer of these groups.

Polyethylene, polypropylene, and polyolefins have no active functional group sites along their length.

Based on the above information, which fabrics would you expect to be dyed by Congo Red or alizarin? Note that alizarin and Congo Red contain numerous hydrogen bonding and polar groups?

ALIZARIN

1. Using tongs, hold a strip of the test material in the beaker of hot alizarin solution for about one minute.

2. Using a paper towel to catch the drips, carry the test material to the sink.

3. Rinse the strip under running water to wash off excess dye, and dry it on a paper towel.

4. When the strip is dry, fasten it on your report page, lining it up with the identification key which shows the order of fabric sections.

5. In a brief paragraph, discuss your results. Why were certain fabrics dyed well, while others were not? How is that related to their intermolecular forces?

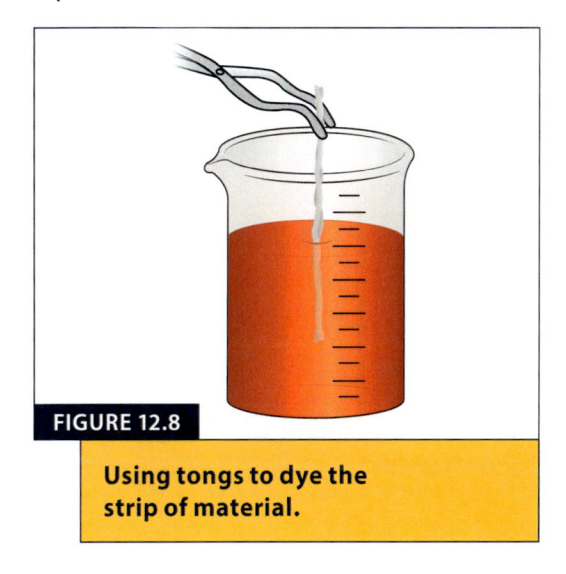

FIGURE 12.8
Using tongs to dye the strip of material.

REPORT FORM

12

NAME: _____

SECTION: _____ DATE: _____

INSTRUCTOR: _____

SOLUTIONS AND SOLUBILITY REPORT FORM

EXPERIMENTAL DATA

Record observations **and** answer questions in space provided.

PART I. UNSATURATED, SATURATED, AND SUPERSATURATED SOLUTIONS

1. Observations with the sodium acetate test.

2. Observations with the potassium chloride tests. Condition:
   a. _____
   b. _____
   c. _____
   d. _____

3. What happens when a few crystals are added to an unsaturated solution? Identify the situation in the experiment where an unsaturated solution existed. Explain.

INTRODUCTORY CHEMISTRY LABORATORY MANUAL

REPORT FORM 12 SOLUTIONS AND SOLUBILITY

4. What happens when a few crystals are added to a supersaturated solution? Identify the situation in the experiment where a supersaturated solution existed. Explain.

5. Was the KCl sample under Condition A unsaturated, saturated, or supersaturated? How do you know?

6. Was the KCl sample under Condition D unsaturated, saturated, or supersaturated? Explain your observation that confirms your answer.

PART II. SOLUBILITY AND STRUCTURE

TABLE 12.1

SUBSTANCE	OBSERVATIONS—BEHAVIOR IN WATER	OBSERVATIONS—BEHAVIOR IN HEXANE	CONCLUSION—PREDICT IF THE SUBSTANCE IS IONIC, POLAR COVALENT, NONPOLAR COVALENT
Water	×		

104 INTRODUCTORY CHEMISTRY LABORATORY MANUAL

REPORT FORM

12

NAME: _____

SECTION: _____   DATE: _____

INSTRUCTOR: _____

SOLUTIONS AND SOLUBILITY REPORT FORM

QUESTIONS

PART I. UNSATURATED, SATURATED, AND SUPERSATURATED SOLUTIONS

1. What would happen if you added more solute to a saturated solution?

2. What would happen if you added more solute to an unsaturated solution?

3. Will a polar solute dissolve in a nonpolar solvent?

4. Is iodine more like a polar water or nonpolar hexane?

5. Based on what happens when cooking oil is added to water, would you expect iodine to dissolve or not dissolve in cooking oil. Why?

REPORT FORM 12 SOLUTIONS AND SOLUBILITY

PART III. DYES AND DYING

TABLE 12.2

ATTACH TEST STRIP HERE	
	Acetate
	Cotton
	Polyamide
	Polyester
	Acrylic
	Silk
	Viscose
	Wool

1. Detailed observations:

2. Conclude as to why did some fabrics dye well with alizarin and others did not.

LABORATORY 13

ACIDS, BASES, AND BUFFERS

OBJECTIVES

- Learn the acid-base characteristics of common substances.
- Use pH paper and a pH meter to characterize the pH of substances.
- Determine the influence of an acid or base on the pH of water.
- Prepare and test a buffer solution.
- Evaluate the acid-base characteristics of exhaled breath.

INTRODUCTION

THE CONCEPT OF pH

The acidity of an aqueous solution depends on the *concentration of hydrogen ions,* [H^+], that it contains. A solution with a large concentration of H^+ is acidic; a solution with a large concentration of hydroxide ion, OH^-, is basic. The concentrations of hydrogen ions can vary by many orders of magnitude. For convenience, the acidity or basicity of a solution is often expressed on the pH scale. Because of the mathematical definition of pH (pH = –log [H^+]), a *low pH represents a high concentration of H^+,* an acidic solution. A solution of pH 1 has an H^+ concentration of 0.1 mole/liter (0.1 M or 10^{-1} M). A solution whose pH is 2 has an [H^+] of 0.01 M (10^{-2} M), one-tenth that of a pH 1 solution. Note that the pH is the negative exponent x when [H^+] is expressed as 10^x Molar. If [H^+] = 10^{-3} M, the pH = 3. Going in reverse, if the pH = 6, [H^+] = 10^{-6} M (or 0.000001 M, an extremely small number). Note that factors of a thousand or even ten million are common in [H^+] changes. A solution with pH 3 is 1000 times more acidic than a solution with pH of 6. The difference is like the difference between walking across the lab room vs walking to High Point or to Los Angeles.

A solution with a pH greater than 7 (concentration of H^+ is less than 10^{-7}) is basic. Because the H^+ concentration is so low, the characteristics of the solution are dominated by the higher concentration of OH^- ions present. At pH 7, the solution contains equal amounts of H^+ and OH^-, and is defined as neutral.

LABORATORY 13 ACIDS, BASES, AND BUFFERS

FIGURE 13.1
The pH scale.

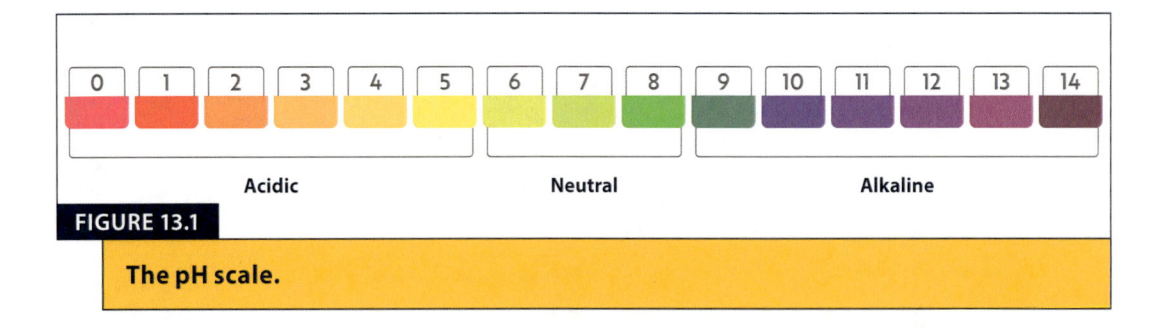

As you may infer from the above information, H⁺ and OH⁻ concentrations are interrelated. Both ions are present in any aqueous solution, and if the amount of one is large (i.e., the H⁺ in an acidic solution), the amount of the other is small. This is represented qualitatively in the diagram below.

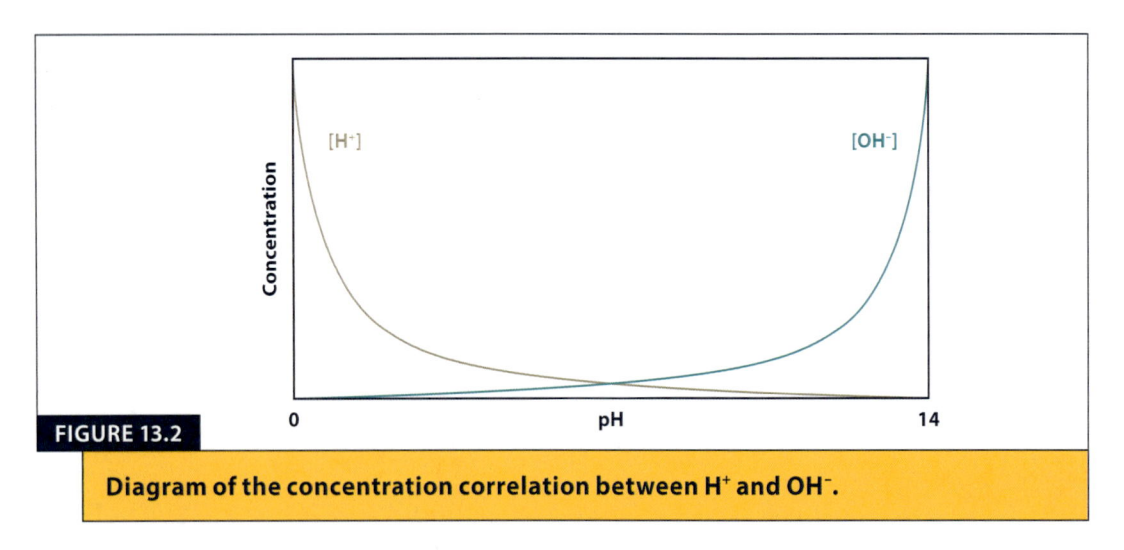

FIGURE 13.2
Diagram of the concentration correlation between H⁺ and OH⁻.

pH is an important factor in many situations outside the research laboratory: in living organisms, the environment, and even to the amateur rose gardener or tropical fish hobbyist. pH can be determined in a variety of ways. Rough estimates can be made with indicators like litmus, phenolphthalein, or Alkacid paper, while more accurate values are obtained using an instrument called a pH meter.

BUFFERS

Often it is desirable to maintain the pH of a chemical system at a relatively constant pH value when small amounts of acid or base are added. This is particularly true in biological systems where changes in pH will have drastic effects on the normal function of the organism. Blood, for example, normally has a pH about 7.4, and the pH must be maintained between 6.8–7.8, or death can result.

Constant pH levels can be achieved with buffers. A **buffer** is a system that resists change in pH when acid or base is added. A buffer usually consists of a salt and a weak acid or a weak base. Blood is a natural buffer.

108 INTRODUCTORY CHEMISTRY LABORATORY MANUAL

LABORATORY 13 ACIDS, BASES, AND BUFFERS

PROCEDURE

The pH probe: Keep the pH electrode immersed in a test solution, deionized water, or the storage buffer **when not in use.** Rinse the electrode off thoroughly, using your wash bottle, before testing each new solution. Also rinse out the beaker thoroughly and be careful not to pour the stirring bar down the drain. The bulb at the end of the pH electrode is extremely fragile. Be careful handling and adjusting the electrode.

PART I. CALIBRATING THE pH METER

The electrode and beaker may carry trace contaminants into the water, and very small amounts of an acid or base can cause the pH of water to be shifted one or two units either side of 7. Rinse glassware thoroughly with tap and then deionized water between measurements.

1. Make sure that the pH probe is attached to the computer and open Logger Pro. A pH reading signifies that the probe is working properly.

2. Calibrate your pH probe using a Two-Point pH Meter Calibration.

    a. Choose "Calibrate" from the Experiment menu and select the sensor you want to calibrate.

    b. Click "Calibrate Now" to begin the calibration process.

    c. Place the sensor in the pH buffer 4 solution found on the back counter.

    d. Enter the known calibration value (4.0) in the Reading 1 box.

    e. Once the **readings in the input field stop** changing significantly, click "Keep" to record the input value. You are only concerned with the *voltage* reading, not the pH reading when you calibrate.

    f. Place the sensor in the pH buffer 7 solution found on the back counter.

    g. Enter the known calibration value (7.0) in the Reading 2 box.

    h. Once the reading in the input field stops changing significantly, click "Keep" to record the input value.

FIGURE 13.3

A pH meter being used to read the pH level of a liquid.

INTRODUCTORY CHEMISTRY LABORATORY MANUAL **109**

LABORATORY 13 ACIDS, BASES, AND BUFFERS

PART II. THE pH OF COMMON SUBSTANCES

On the side shelf you will find the following substances with sample applicators: ammonia, shampoo, ivory, lemon juice, Coke, and/or vinegar.

1. Determine the pH of the provided common household substances (liquids only) with the pH meter by filling a large test tube with just enough material to cover the electrode.

2. Use Alkacid pH test paper to find the approximate pH of each of the substances provided. Estimate to the nearest whole pH unit. (One piece of pH paper can be used for three or four tests if you place a drop of the solution on the paper using your glass stirring rod.)

FIGURE 13.4 Common household liquids.

Note: On the color chart pH 4 is labeled "very strong acid." A better description is perhaps "Moderately acidic." pH 2 is "strong acid" and "very strong acid" would fall in the range of pH 1 to pH 0.

PART III. EFFECT OF ACID AND BASE ON pH OF WATER

1. Place 50 mL of deionized water in a beaker. Immerse the electrode and record the initial pH reading. It will likely read higher or lower than 7 and that is ok.

2. Use the dropper bottle of the 1 M HCl provided and expel all of the solution that is in the dropper. Just touch the tip of the dropper to the water. Record the pH. Mix carefully by stirring with the electrode.

3. Repeat for the addition of 1 drop, 2 more drops, 5 more drops, and a dropper full of HCl. Place the pH probe in deionized water while you clean up and prepare for the next procedure.

4. Repeat the above procedure with 50 mL of deionized water and the dropwise addition of 1 M NaOH.

PART IV. EFFECT OF ACID AND BASE ON THE pH OF AN ACETATE MIXTURE

1. Mix 40 mL of 0.5 M acetic acid ($HC_2H_3O_2$) and 4.0 mL of 0.5 M sodium acetate ($NaC_2H_3O_2$) in a clean beaker and record the pH of the solution.

2. Add 1 M HCl dropwise as in Part III, recording the pH after each addition.

3. Repeat the above procedure on a fresh mixture of $HC_2H_3O_2/NaC_2H_3O_2$, using 1 M NaOH in place of the HCl.

   Discuss your results on the Report Form.

PART V. EFFECT OF ACID AND BASE ON THE pH OF ALKA-SELTZER®

1. Place one tablet of Alka-Seltzer in 100 mL of deionized water and stir to dissolve it completely. Transfer half of the solution to a beaker and determine the pH.

2. Then add 1 M HCl dropwise, recording the pH as you did earlier.

3. Repeat the procedure with the other half of the Alka-Seltzer solution and 1M NaOH solution.

PART VI. EFFECT OF CO_2

1. Put 50 mL of deionized water into a clean beaker and record the pH. Add 1 drop of 0.1 M NaOH; what is the pH of the solution now? Add 3 drops of bromothymol blue indicator and note the color.

2. Obtain a straw and blow into the solution with one good exhaled breath, observing the effect of the CO_2 from your breath on the pH as you blow. Repeat for several more breaths, pausing to record the pH each time. Look for a point where the pH changes very little. Discard the straw.

REPORT FORM 13

NAME: _____

SECTION: _____ DATE: _____

INSTRUCTOR: _____

ACIDS, BASES, AND BUFFERS REPORT FORM

EXPERIMENTAL DATA

PART I & II. ACIDITY OF SOME COMMON SUBSTANCES

TABLE 13.1

SUBSTANCE	pH METER READING	ESTIMATED pH pH PAPER

REPORT FORM 13 ACIDS, BASES, AND BUFFERS

1. Comment on the values obtained in pH measurements of common substances. Were any values surprising? Can you classify the results in any way? Compare the use of pH paper and a pH meter.

PART III. EFFECT OF ACID AND BASE ON THE pH OF WATER

TABLE 13.2

		PURE WATER	1/10 DROP	1 DROP	2 MORE DROPS	5 MORE DROPS	FULL DROPPER
Addition of 1 M HCl	pH						
Addition of 1 M NaOH	pH						

PART IV. EFFECT OF ACID AND BASE ON THE pH OF ACETATE MIXTURE

TABLE 13.3

		ACETATE MIXTURE	1/10 DROP	1 DROP	2 MORE DROPS	5 MORE DROPS	FULL DROPPER
Addition of 1 M HCl	pH						
Addition of 1 M NaOH	pH						

REPORT FORM

13

NAME: _____

SECTION: _____ DATE: _____

INSTRUCTOR: _____

ACIDS, BASES, AND BUFFERS REPORT FORM

PART V. EFFECT OF ACID AND BASE ON THE pH OF ALKA-SELTZER

TABLE 13.4

		ALKA-SELTZER	1/10 DROP	1 DROP	2 MORE DROPS	5 MORE DROPS	FULL DROPPER
Addition of 1 M HCl	pH						
Addition of 1 M NaOH	pH						

PART VI. EFFECT OF CO_2

1. a. pH of deionized water: _____

   b. pH after addition of 1 drop of 0.1 M NaOH: _____

   c. pH after blowing in CO_2: _____

   d. Further pH changes: _____

INTRODUCTORY CHEMISTRY LABORATORY MANUAL 115

REPORT FORM 13 ACIDS, BASES, AND BUFFERS

RESULTS AND DISCUSSION

2. a. In Part III, what was observed when HCl was added to deionized water? When NaOH was added? Does the addition of only a drop or two of acid or base to water cause a significant change in pH?

   b. Consider that each 1 pH unit change represents a factor of 10. How many more times acidic is the water after 1 drop of acid was added to the water than before any acid was added?

3. Compare Part III with Part IV. Describe which system shows buffering and explain why.

4. Compare the effects of Alka-Seltzer and the acetate system. Which exhibits buffering? What was the main difference between them?

REPORT FORM 13

NAME: _____

SECTION: _____ DATE: _____

INSTRUCTOR: _____

ACIDS, BASES, AND BUFFERS REPORT FORM

5. Discuss the effect of the CO_2 on the solution in Part VI. Based on its behavior in the experiment, is CO_2 acidic or basic? Did the pH continue to change, or did it reach a constant value? Why?

6. You may have noticed that the deionized water in the lab is not pH 7 as you might have expected it to be. Based on this experiment, give one reason for the pH of water being different than 7.

INTRODUCTORY CHEMISTRY LABORATORY MANUAL **117**

LABORATORY 14

DETERMINING THE PURITY OF ASPIRIN BY TITRATION

OBJECTIVES

- Use volumetric equipment (pipet and buret) to accurately measure volumes of solutions.
- Use titration technique to accurately determine the amount of a chemical substance in a sample.
- Calculate percent purity from analysis of a sample.
- Use multiple trials to evaluate the reproducibility of an analysis.

INTRODUCTION

The aspirin which you previously synthesized according to Figure 14.1 is probably not pure, despite your best efforts. The most likely impurities are acids—either salicylic acid from unreacted starting material or acetic acid, a byproduct of synthesis. Even commercially prepared aspirin tablets are not 100 percent acetylsalicylic acid. Most aspirin tablets contain a small amount of binder which helps prevent the tablets from crumbling. The binder is chemically inert and was intentionally added by the manufacturer, but its presence means that aspirin tablets do not have 100 percent purity.

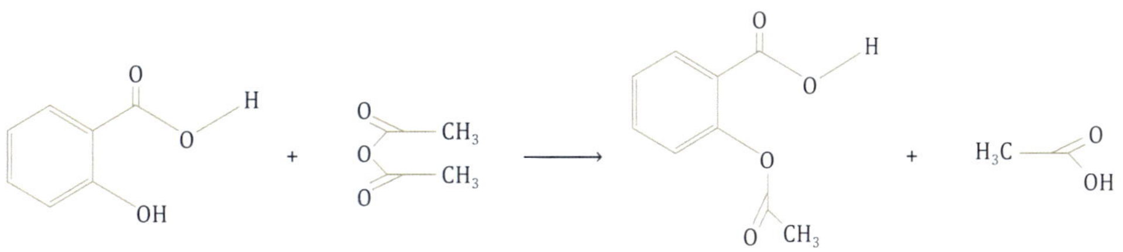

FIGURE 14.1

Aspirin Synthesis.

INTRODUCTORY CHEMISTRY LABORATORY MANUAL 119

Moreover, moisture can hydrolyze acetylsalicylic acid. Thus, aspirin that is not kept dry can decompose. Acetic acid is the hydrolysis product formed by the reaction of water with acetylsalicylic acid as shown in Figure 14.2.

FIGURE 14.2

Aspirin Hydrolysis.

In this experiment, we will analyze your synthesized aspirin and commercial aspirin for purity using qualitative and quantitative methods. Qualitatively, you can test for the presence of a phenol functional group in an aspirin sample with the addition of ferric chloride solution. The ferric ion will form a violet compound if a phenol is present; the deeper the violet color, the more impurities that are present. Quantitatively, the purity of an aspirin sample can be determined by a simple acid-base titration.

A titration is a common procedure used in the chemistry laboratory that is based on acid-base neutralization theory. It is centered on the concept that in a neutralization reaction, the number of moles of base exactly equals the number of moles of acid—and vice versa according to the equation below. When performing a titration, a base of known concentration is added to a known amount of acid of unknown concentration until the acid is exactly neutralized.

FIGURE 14.3

A Titration Setup.

$$HA(aq) \text{ acid} + NaOH(aq) \text{ base} \rightarrow H_2O(l) + NaA(aq) \text{ salt.}$$

Table 14.1 compares the chemical properties of both salicylic acid and aspirin. Note that salicylic acid has a lower molar mass. This means that if we titrate the same mass of each substance it will take more NaOH to reach the endpoint with salicylic acid because there are more molecules. Thus, if our aspirin is contaminated with salicylic acid, it will require more NaOH to reach the endpoint than if our sample was 100% pure aspirin; a mixture of the two acids would require an intermediate amount of base.

TABLE 14.1

COMPOUND	SALICYLIC ACID	ASPIRIN
Formula:	$C_7H_6O_3$	$C_9H_8O_4$
Molar Mass:	138.12	180.15
Melting point:	158–160 °C	140–142 °C
Solubility (g/100 mL)	0.18	0.25

You will calculate the % purity of the aspirin samples using the equation below. For the commercial aspirin, the theoretical moles of aspirin can be calculated from the amount of active ingredient stated on the bottle's label. It is assumed that your entire sample is aspirin, so a mass to mole conversion can be used to determine the moles of aspirin in your sample. The actual number of moles is calculated from the amount of NaOH used to titrate the sample.

% Purity = (actual moles of aspirin/theoretical moles of aspirin) × 100%.

PROCEDURE

PART I. FERRIC CHLORIDE TEST

1. Dissolve a few crystals of commercial aspirin, salicylic acid, and your synthesized aspirin in 3 separate test tubes with 5 drops of 95% ethanol.
2. Increase the volume of your sample by adding 5 drops of water and mix well.
3. Add 2 drops of 1% Iron (III) Chloride solution to each, mix well, and observe the color of each sample.

PART II. SYNTHESIZED ASPIRIN TITRATION

1. Weigh out approximately 0.4 grams of synthesized aspirin into a beaker. Record its exact mass.
2. Add about 10 mL of 95% ethanol and swirl to mix well to dissolve the acid in your sample.
3. Add about 25 mL of DI water and mix well.

4. Add 2 drops of phenolphthalein indicator and mix well.
5. Fill a clean 50 mL burette with standardized NaOH.
6. Read the initial volume on the burette to 2 decimal places.
7. Slowly, add the NaOH solution to the flask with continuous swirling until the endpoint is reached (persistent light pink color).
8. Record the final burette reading.
9. Repeat the titration until two NaOH volume values are within +/− 0.10 mL of each other (this may require more than two titrations).

PART III. COMMERCIAL ASPIRIN TITRATION

1. Place an entire aspirin tablet into a beaker and add 3 drops of water on top. Allow the water to soak into the tablet for a few minutes and then gentle crush the aspirin tablet into a powder using a stirring rod.
2. Repeat Steps 2–9 above and calculate the percent purity of commercial aspirin.